Horticultural Crops

Edited by Hugues Kossi Baimey, Noureddine Hamamouch and Yao Adjiguita Kolombia

Published in London, United Kingdom

IntechOpen

Supporting open minds since 2005

Horticultural Crops
http://dx.doi.org/10.5772/intechopen.73771
Edited by Hugues Kossi Baimey, Noureddine Hamamouch and Yao Adjiguita Kolombia

Contributors
Yelitza Colmenarez, Carlos Vásquez, Marcus Vinicius Sampaio, Simone Mundstock Jahnke, Natália Corniani, Pazhayaveetil Kuttanpillai Chandrasekhara Pillai, Sanal Chalil Viswanath, Thoduvayil Karunakaran Hrideek, Aviyan Hari Jiji, Samar S. Ibrahim, Vladimir Volynkin, Alla Polulyakh, Svetlana Levchenko, Irina Vasylyk, Vladimir Likhovskoy, Mesfin Kebede Gessese, Junyang Song, Ning Zhang, Fatin Nadiah Jamil, Chu-Nie Tang, Noor Baity Saidi, Nadiya Akmal Baharum, Kok-Song Lai, Juan Carlos Álvarez-Hernández, Waldir Cintra de Jesus Junior, Edson Luiz Furtado, Willian Bucker Moraes, Breno Benvindo Dos Anjos, Lilianne Gomes Da Silva, Muhammad Ibrahim Khaskheli

First published in London, United Kingdom, 2020 by IntechOpen
IntechOpen is the global imprint of INTECHOPEN LIMITED, registered in England and Wales, registration number: 11086078, 7th floor, 10 Lower Thames Street, London, EC3R 6AF, United Kingdom
Printed in Croatia

British Library Cataloguing-in-Publication Data
A catalogue record for this book is available from the British Library

Additional hard and PDF copies can be obtained from orders@intechopen.com

Horticultural Crops
Edited by Hugues Kossi Baimey, Noureddine Hamamouch and Yao Adjiguita Kolombia
p. cm.
Print ISBN 978-1-83880-421-3
Online ISBN 978-1-83880-422-0
eBook (PDF) ISBN 978-1-83880-437-4

For EU product safety concerns:
IN TECH d.o.o., Prolaz Marije Krucifikse Kozulić 3, 51000 Rijeka, Croatia, info@intechopen.com or visit our website at intechopen.com.

Meet the editors

Dr. Hugues Kossi Baimey, Associate Professor, graduated with his MSc in nematology from Ghent University, Belgium (2000) and obtained a PhD at Pretoria University, South Africa (2006) on yam nematodes. He is currently a lecturer (2009-present) in the Faculty of Agronomy, University of Parakou, Benin where he gives eight courses including plant-parasitic and entomopathogenic nematology. His research interests are primarily centered on the management of agricultural insect pest populations using entomopathogenic nematodes and the control of plant-parasitic nematodes associated with vegetable, root, tuber, and cereal crops. He is a past Director of the Higher National School of Agricultural Sciences and Techniques, University of Parakou, Benin (2012-2016) and Director of international projects funded by Nuffic (2012-2016) and DANIDA (2015-2016). He has been Principal Investigator of international projects funded by the VLIR-UOS (2010-2014), the Bill & Melinda Gates Foundation (2015-2019), and the National Academy of Science/USAID (currently). Hehas also received funds from Pro-Sol/GIZ (2019), the GRiSP (2012-2014), BEST (2010-2014), Agropolis-Foundation-Capes (2011-2013), and TWAS (2008-2009) for nematology research. Dr. Kossi Baimey has worked for IITA, Cotonou Station (1995-2009 and 2016-2018), University of Alicante, Spain (2001); AfricaRice Cotonou Station (2009-2011) and has had collaborations with IRD Montpellier, France (2010-2016); Ghent University, Belgium (2010-present); Oxfam Quebec NGO, Benin (2010); GERES NGO, Benin (2012-2014); eNema, Germany (2015-2019); CSIR-CRI and CSIR-SARI, Ghana (2015-2019). He has published 62 articles in peer-reviewed journals and 47 conference abstracts. He is a co-author of two books: "Searching for better methodologies for successful control of termites using entomopathogenic nematodes" (In: \Nematology - Concepts, Diagnosis and Control\, IntechOpen, 2017) and "Integrated pest management in vegetable production: A guide for extension workers in West Africa" (IITA, 2010). He has supervised two PhD students and around 40 Master and 35 BSc students.

Prof. Noureddine Hamamouch obtained his PhD degree in Molecular Cell Biology and Biotechnology from Virginia Polytechnic Institute and State University in Virginia USA, and worked as a postdoctoral researcher at the University of Pittsburgh Medical Center and at North Carolina State University. Prof. Hamamouch uses molecular tools to examine plant-pathogen interactions and genetically engineer plant resistance to pathogens. He has been selected by the United States Department of Agriculture as a Biotechnology Expert to the Biotechnology Center in Morocco. Prof. Hamamouch was a Lecturer at Longwood University and at Central Virginia Community College, both in Virginia, USA. He is currently an Associate Professor at the University of Mohamed V, in Rabat, Morocco. Prof. Hamamouch has given lectures at different universities in USA and Europe (Belgium, Portugal, and Spain), published 15 scientific articles in high impact peer-reviewed international journals including Nature Biotechnology, Journal of Experimental Botany and Phytopathology, and has received numerous grants and scholarships including the Erasmus Mundus Grant.

Dr. Yao Adjiguita Kolombia obtained his PhD in Biology from Ghent University, Ghent, Belgium. Dr. Kolombia is an agricultural engineer with research experience in the biological control of weeds, nematode taxonomy, and management. Dr. Kolombia is currently the banana pathologist at the International Institute of Tropical Agriculture (IITA). Dr. Kolombia received several grants among which, the Erasmus Mundus Grant, the Special Research Fund (BOF), and the Yam improvement for income and food security in West Africa (YIIFSWA). He has published several papers in impact factor journals.

Contents

Preface

Horticultural crops are important sources of proteins, minerals, carbohydrates, organic acids, and vitamins for human nutrition. They are some of the most important components of a healthy diet. But the successful cultivation of these crops necessitates sufficient knowledge, skills, and technologies from seed selection and planting to harvesting and storage. The use of quality material for planting is a prerequisite for successful horticulture. Also, during their development phases, crop plants are exposed to several biotic stresses (disease-causing agents and pests) and abiotic stresses that lead to a series of physiological, morphological, biochemical, and molecular changes, both at harvest and during storage. This causes important horticultural crop yield losses. The management of those stresses is therefore important to ensure good crop yields and in such a context is viewed as an integral part of crop production. To collect precise information on the type of stresses that undermine the crops during their development or in stores, comprehensive diagnostic surveys are conducted and when causal agents are identified, their management options are selected and applied. Several methods are used in this regard, including: genetics and crop improvement and chemical, botanical, and biological control methods.

This book is aimed at students and researchers. It is divided into three main sections. Section 1 presents information on the physiology of some horticultural crops such as Malabar Bauhinia (*Bauhinia malabarica* Roxb) and wild multi-flower orchid (*Cymbidium faberi*) (Chapters 1 and 2, respectively). Section 2 provides examples of the use of genetics and the improvement of coffee (Chapter 3), tomato, watermelon, papaya, and some other horticultural crops (Chapter 4) and grapes (Chapter 5) for the management of viral, fungal, bacterial, and nematode diseases. Section 3 describes the epidemiology and management of South American leaf blight (SALB) caused by the fungus *Microcyclus ulei* associated with rubber (*Hevea* spp.) cultivation in Brazil, fusarium wilt caused by *Fusarium oxysporum* in bananas (*Musa* spp.), and several infectious diseases caused by many phytopathogens in mango (*Mangifera indica* L.) in Chapters 6, 7, and 8, respectively. The last two chapters of Section 3 present information on the use of parasitoids and essential oil nanoformulations for the control of insect pests associated with horticultural crops. The editors record their thanks to all authors of the different chapters of this book. They are especially grateful to Ms Sandra Maljavac for their valuable contribution during the editing processes leading to book publication.

Hugues Kossi Baimey
University of Parakou,
Benin

Noureddine Hamamouch
University Sultan Moulay Slimane,
Morocco

Yao Adjiguita Kolombia
International Institute of Tropical Agriculture,
Ibadan Station, Nigeria

Section 1

Plant Physiology

Chapter 1

Seed Characteristics and Germination Behaviour of *Bauhinia malabarica* Roxb.

Pazhayaveetil Kuttanpillai Chandrasekhara Pillai, Sanal Chalil Viswanath, Thoduvayil Karunakaran Hrideek and Aviyan Hari Jiji

Abstract

Malabar *Bauhinia* (*Bauhinia malabarica*) is a native ornamental species belonging to the family Fabaceae, distributed throughout India in semievergreen and moist deciduous forests and in gardens. Information regarding seed characteristics and seed handling of the species is meagre. This study describes seed characteristics, germination behaviour and pretreatment for enhancing seed germination of *B. malabarica*. Treating the seeds with concentrated sulphuric acid for 30 min reduced germination duration up to 6 days and increased germination rate up to 100% against 22 days of germination duration and 10% germination in seeds without treatment. The results of this study are helpful for conservation and nursery practices of *B. malabarica*.

Keywords: maturity index, seed germination, germination value, pretreatment, sulphuric acid

1. Introduction

Bauhinia malabarica Roxb., commonly known as Malabar *Bauhinia*/mountain ebony, is a deciduous tree belonging to the family of Fabaceae, distributed all over the Indo-Malayan region. The species is found throughout Indian subcontinent and popularised in different vernacular names as Arampuli (Malayalam), Malayathi (Tamil), Basavana pada (Kannada), Amli (Hindi), etc. In native range, the species is distributed in the semievergreen and moist deciduous forests up to 600 m. It is used as an ornamental plant in homesteads and gardens of the native range.

Studies of the genus *Bauhinia* is very limited compared with other genus like *Caesalpinia*, *Cassia*, *Tamarindus*, etc., under the family Fabaceae. Primary documentation of the genus *Bauhinia* in India was conducted during the eighteenth century [1]. Neotypification of the species *Bauhinia malabarica* was recently done by the Central National Herbarium of Botanical Survey of India [2]. Micromorphological characteristics of selected species of the genus *Bauhinia* were studied earlier [3]. Variation in size and structure of selected species of *Bauhinia* was examined and recorded by different scholars [4, 5]. Analysis on mineral elements and nutritional and anti-nutritional contents in the seeds of *B. monandra* was conducted in an earlier study [6]. Phytochemical structure like polysaccharide and proteins in seeds of *Bauhinia* was investigated earlier [7, 8]. Antibacterial nature of *B. acuminata* was reported in a previous study [9]. It was reported that seed oil of *Bauhinia* is a novel substance for the production of sophorolipids [10].

A few investigations were done on *B. malabarica* on foliar micromorphology, natural regeneration, chemical composition, phytochemical analysis, antimalarial activities and anti-oxidant potential [3, 11–15]. Studies related to seed germination in the genus *Bauhinia* was conducted only on selected species like *B. rufescens*, *B. thonningii*, etc. [16–18]. However, information on seed characteristics and germination parameters of *B. malabarica* is limited. The present study was done to fill the above-mentioned gaps.

2. Materials and methods

Periodical observations were carried out on mother plants of *B. malabarica* to identify optimum maturity index for seed collection. Mature pods (fruits) of *B. malabarica* were collected (March–April, 2016) from Peechi-Vazhani Wildlife Sanctuary (10°31′48″N; 76°20′50″E) in Thrissur District, Kerala, India. The pods were dried under sunlight (35–38°C) for 2 days, and seeds were extracted by splitting the pods. Extracted seeds were dried in shade, cleaned and stored in airtight containers. Seed characteristics such as morphology, moisture content and germination were evaluated. High constant temperature oven-dry method was used to determine seed moisture content [19]. The seeds were dried in hot-air oven for 1 h at 130°C. Seed moisture content (MC %) was estimated according to the formula of ISTA.

$$MC\% = \frac{fresh\ weight\ of\ seed - oven\ dry\ weight\ of\ seed}{fresh\ weight\ of\ seed} \times 100 \quad (1)$$

Seeds were subjected to different pre-sowing treatments to enhance seed germination and reduce the germination period. The following were the pretreatments applied for the study:

T1: control (no pre-sowing treatment).
T2: soaked in water for 24 h.
T3: soaked in water for 48 h.
T4: soaked in hot water for 2 min.
T5: soaked in hot water for 5 min.
T6: soaked in hot water for 2 min + soaked in water for 24 h.
T7: soaked in hot water for 5 min + soaked in water for 24 h.
T8: soaked in GA_3 (500 ppm) for 2 h.
T9: soaked in GA_3 (1000 ppm) for 2 h.
T10: acid treatment (conc. H_2SO_4) for 10 min.
T11: acid treatment (conc. H_2SO_4) for 20 min.
T12: acid treatment (conc. H_2SO_4) for 30 min.

Tap water (≈35°C) was used in T2 and T3 treatments, whereas in the treatments T4–T7, hot water (85°C) was used. Different concentrations of gibberellic acid (GA_3)/gibberellin A3 (chemical formula: $C_{19}H_{22}O_6$) were used in T8 and T9 treatments. Concentrated sulphuric acid (98%) was used in the treatments T10–T12.

Seeds (n = 100 in 4 replications) were sown in germination trays having a size of 25 × 20 × 5 cm filled with vermiculite and kept in germination room (30 ± 2°C and 90% RH) under laboratory condition. Randomised block design was adopted for the experiment. Data on seed germination were recorded starting from seed germination till culmination and computed germination-related parameters. Germination initial time (GIT), germination percentage (GP), germination duration (GD), mean germination time (MGT), mean daily germination (MDG), germination energy (GE), peak value (PV) and germination value (GV) were calculated [20–24].

Germination-related parameters were determined as follows:

$$\text{Germination initial time (GIT)} = \text{Dg–Ds} \tag{2}$$

where Dg = first germination day and Ds = seed sowing day.

$$\text{Germination percentage (GP)} = (\text{G/T}) \times 100 \tag{3}$$

where G = no. of germinated seeds and T = no. of seeds sown.

$$\text{Germination duration (GD)} = \text{Gf–Gi} \tag{4}$$

where Gf = final day of germination and Gi = initial day of germination.

$$\text{Mean germination time (MGT)} = (\text{Gt} \times \text{Dt})/\text{G} \tag{5}$$

where Gt = no. of germinated seeds at day-t, Dt = no. of days at 't' from the day of sowing and G = total no. of germinated seeds.

$$\text{Mean daily germination (MDG)} = \text{GP/Gd} \tag{6}$$

where GP = germination percentage and Gd = no. of days to complete germination.

$$\text{Germination energy (GE)} = \text{X1/Y1} + (\text{X2} - \text{X1})/\text{Y2} + \ldots\ldots. + (\text{Xn} - \text{Xn} - 1)/\text{Yn} \tag{7}$$

where Xn = no. of germinants on the nth counting date and Yn = no. of days from sowing to the nth count.

$$\text{Peak value (PV)} = \text{Highest number of seeds germinated/no.of days required to the peak germination} \tag{8}$$

$$\text{Germination value (GV)} = \text{PV} \times \text{MDG} \tag{9}$$

where PV = peak value and MDG = mean daily germination.

2.1 Statistical analysis

Each trait was analysed using mean values under the various pretreatments. The variation on mean values between these treatments were performed through analysis of variance (ANOVA) done by statistical software SPSS version 22.

3. Results

3.1 Seed weight and moisture content

The study recorded 7092 ± 50 seeds per kilogram. Moisture content (MC %) of fresh seeds was 5.35%.

3.2 Maturity index

The optimum maturity indices for seed collection of *B. malabarica* were presented in **Table 1**. The maturity indices identified for determining optimum period for seed collection were colour of pods (yellowish-green turned to blackish-green), leaf number (minimum number of leaves), dehydration (pods become dehydrated) and hardness (pods and seeds become hardest).

3.3 Pod/seed characteristics

Colour, shape, size, type, weight, number of seeds per pod and per kg, type of germination, etc. are presented in **Table 2**.

3.4 Seed germination

Figure 1 depicts germination pattern of seeds under various pretreatments. Seed germination among treatments was significantly different (**Table 3**).

Germination-related parameters under different pretreatments such as germination initial time, germination percentage, germination duration, mean germination time, mean daily germination, germination energy, peak value and germination value are given in **Table 4**.

Character	Variable	Nature
Fruit	Colour	Yellowish-green turned to blackish-green
	Water content	Very less
	Hardness	Hardened
Seed	Colour	Dark brown
	Hardness	Hardened
Leaf	Number	Minimum

Table 1.
Pod/seed maturity indices of B. malabarica.

Character	Variable	Nature
Fruit	Type	Pod
	Colour (young)	Green
	Colour (mature)	Blackish brown
	Size (cm)	15.25 ± 4.75 × 2.15 ± 0.25
	Wall type	Dry, semihard
Seed	No. of seeds/fruit	10 ± 2
	Colour	Dark brown
	Shape	Broad elliptic/oblong
	Size (mm)	.02 ± 1.11 × 6.14 ± 0.88
	Wall type	Dry, hard
	Weight	7092 ± 50
Germination	Type	Epigeal

Table 2.
Seed characteristics of B. malabarica.

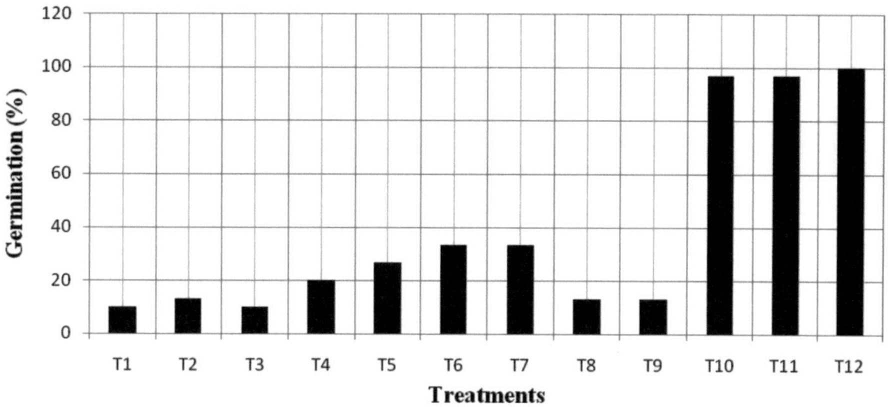

Figure 1.
Seed germination pattern under different treatments. Note: T1, control; T2, soaked in water for 24 h; T3, soaked in water for 48 h; T4, soaked in hot water for 2 min; T5, soaked in hot water for 5 min; T6, soaked in hot water for 2 min + soaked in water for 24 h; T7, soaked in hot water for 5 min + soaked in water for 24 h; T8, soaked in gibberellic acid (GA_3—500 ppm) for 2 h; T9, soaked in gibberellic acid (GA_3—1000 ppm) for 2 h; T10, acid treatment (conc. H_2SO_4) for 10 min; T11, acid treatment (conc. H_2SO_4) for 20 min; T12, acid treatment (conc. H_2SO_4) for 30 min.

Source	Sum of squares	Degrees of freedom (df)	Mean square	F-value	Sig.
Between groups	4332.775	11	393.889	29.804	0.005
Within groups	2669.618	202	13.216		
Total	7002.393	213			

Table 3.
ANOVA table—level of significance on mean values of seed germination under different pretreatments.

Germination Initial Time (GIT): Germination initial time was 2 days after sowing in the treatments T4, T10, T11 and T12, whereas it was 5 days after sowing in T1, T2, T5, T8 and T9 treatments.

Germination Percentage (GP): 100% germination was achieved in T12 treatment (acid scarification for 30 minutes) followed by 97% in T11 and T10 (acid scarification for 20 and 10 minutes). However, 10% was noticed in TI (no treatment) and T3 (soaked in water for 48 h) followed by 13.3% in T2 (soaked in water for 24 h), T8 (500 ppm GA_3) and T9 (1000 ppm GA_3). Hot water treatments followed by water soaking (T6 and T7) exhibited about 33% germination, which was better than mere hot water treatment (T4, 20%, and T5, 27%).

Germination Duration (GD): Seed germination started 2 days after sowing and completed in 28 days. The least GD noticed in treatment T12 (06 days) followed by T11 (09 days) and T10 (11 days) and the highest in T2 (26 days).

Mean Germination Time (MGT): Mean germination time was more in T12 treatment (21.67) where the treatment gave 100% germination, whereas the least value was in seeds without any treatment (T1).

Mean Daily Germination (MDG): Mean daily germination was highest in the treatment T12 (16.67), and the lowest value was in T3 treatment (0.43).

Germination Energy (GE): Germination energy in various pretreatments showed that the highest GE was noticed in T12 (6.85), followed by T11 (5.81) and T10 (4.48). The least GE is in T3 (0.29) followed by T1 (0.30).

Peak Value (PV): Peak value was highest in T12 treatment (12.50), and least value (0.56) was noticed in treatments T1, T3, T8 and T9.

Germination Value (GV): The highest germination value was also observed in T12 (208.36), followed by T11 (116.31) and T10 (58.63).

Treatment	GIT	GP	GD	MGT	MDG	GE	PV	GV
T1	5	10	22	1.77	0.45	0.30	0.56	0.25
T2	5	13.3	26	2.23	0.51	0.42	1.11	0.57
T3	4	10	23	1.83	0.43	0.29	0.56	0.24
T4	2	20	25	3.28	0.80	0.59	1.11	0.89
T5	5	26.7	25	4.08	1.07	0.81	1.11	1.19
T6	4	33.3	25	6.56	1.33	0.78	0.67	0.89
T7	4	33.3	24	5.54	1.39	0.92	1.11	1.54
T8	5	13.3	21	2.57	0.63	0.37	0.56	0.35
T9	5	13.3	21	2.57	0.63	0.37	0.56	0.35
T10	2	96.7	11	18.27	8.79	4.48	6.67	58.63
T11	2	96.7	9	17.56	10.74	5.81	10.83	116.31
T12	2	100	6	21.67	16.67	6.85	12.50	208.36

Note: *T1, control; T2, soaked in water for 24 h; T3, soaked in water for 48 h; T4 soaked in hot water for 2 min; T5 soaked in hot water for 5 min; T6, soaked in hot water for 2 min + soaked in water for 24 h; T7, soaked in hot water for 5 min + soaked in water for 24 h; T8, soaked in* GA_3 *(500 ppm) for 2 h; T9, soaked in* GA_3 *(1000 ppm) for 2 h; T10, acid treatment (conc.* H_2SO_4*) for 10 min; T11, acid treatment (conc.* H_2SO_4*) for 20 min; T12, acid treatment (conc.* H_2SO_4*) for 30 min; GIT, germination initial time; GP, germination percentage; GD, germination duration; MGT, mean germination time; MDG, mean daily germination; GE, germination energy; PV, peak value; GV, germination value.*

Table 4.
Seed germination-related parameters under different treatments.

4. Discussion

Maturity indices help to collect seeds with maximum viable seeds. The optimum maturity indices of *B. malabarica* identified in the present study were the colour of pod turned from yellowish-green to blackish-brown, minimum number of leaves, pods become dehydrated and the pods and seeds become hardest. A previous study reported that seed germination of *Albizia lebbeck* significantly influenced by date of pod collection [25]. Seed weight of *B. malabarica* in the present study showed 7092 ± 50 seeds per kilogram. However, in an earlier report, it is 1100–2600 seeds per kilogram [26]. Seed size is usually related with its vigour and a measure of potential performance; hence, seed weight is significant. Information regarding seed weight and moisture content is helpful for nursery practices and research-oriented studies and also an updating of the earlier information regarding seed weight and moisture content and seed characteristics.

The present study indicated that the highest germination was recorded in acid treatments (acid scarification for 30, 20 and 10 min). All other treatments exhibited poor performance in germination. Germination initiation period was minimum in the treatments T4, T10, T11 and T12, whereas maximum was in T1, T2, T5, T8 and T9 treatments. The lowest germination initial time shows speedy initiation of germination among pretreatments.

Estimation of germination percentage is the best tool to explain seed viability of a particular lot. Seeds treated with concentrated sulphuric acid for 30 minutes (T12) resulted in very high germination rate (100%) than that of other treatments. The study also revealed that the hormonal treatment had no significant role on seed germination of *B. malabarica*. A similar result was reported in an earlier study on *B. rufescens* after acid treatment [17]. Better performance in acid scarification on seed germination was reported in seeds of many species having hard seed coat [27–29]. Germination percent is useful for computing seed requisite for desired number of plants. Pretreatment with high germination value indicates the germination power of seeds.

Germination duration (GD) is helpful to understand the duration required for completing the process of germination. Germination duration in the present study varied with treatments (06–26 days). The study showed that the seeds scarified with concentrated sulphuric acid for 30 min helped to reduce germination period into 6 days compared to 26 days in seeds soaked in water for 24 h. The lowest GD shows the minimum period required to complete germination among pretreatments. Mean germination time (MGT) is the indicative of emergence performance of seed lots. Mean germination time (MGT) was highest in seeds treated with concentrated sulphuric acid for 30 min (21.67) where the treatment gave 100% germination, whereas the least value was in seeds without any treatment (T1). Similarly, mean daily germination (MDG), germination energy (GE), peak value (PV) and germination values were also highest in the treatment T12.

Mean germination time and mean daily germination are used as a gauge of the rate and time spread of germination. High MGT and MDG values indicate high germinability of seed lots due to pretreatments. Peak value indicates the maximum germination rate in a particular day, and germination value is the expected seedlings in the field or nursery. Germination energy and germination value are the easier way to understand the rate of germination and period of germination. Highest GE and GV show the enhanced germination and reduced duration.

The study resulted in scarification of seeds by concentrated sulphuric acid for 30 min which was the best pre-sowing treatments for enhancing seed germination and reduce germination period in *B. malabarica*. Previous studies showed that acid scarification is the best pretreatment to improve germination of seeds with hard seed coat [27–29]. Similarly, parameters like germination percentage, germination energy, mean germination time, mean daily germination, peak value and germination value were also highest in the acid treatment for 30 min compared to other treatments. High values of parameters indicate the germination potential of the seeds. Germination energy/germination value is the tool for indexing the speed and completeness of seed germination [22]. High germination energy and germination value show the effects of pretreatment on seed germination. Period of seed germination in *B. malabarica* is reported in an earlier study with 6–30 days and hot water treatment for 1 min followed by soaking in cold water for 24 hours as the best pretreatment [26]. Similarly, a high rate of germination observed in seeds of *Hippophae salicifolia* treated with thiourea [30]. The Forest Research Institute, India, reported only a low rate of seed germination (14–18%) in *B. malabarica* [31]. However, the present study revealed that the concentrated sulphuric acid treatment for 30 min shall reduce germination period (6 days) and increase germination rate (100%). An earlier study on *Macaranga peltata* revealed that combination of concentrated sulphuric acid and gibberellic acid resulted in improved germination rate [32].

5. Conclusions

The study gave update to seed characteristics and germination behaviour of *B. malabarica*. The investigation has documented maturity indices for determining optimum seed collection period of the species *B. malabarica*, and the data is very helpful for further seed biological studies and experiments on the species. The study suggested optimum period for collection of seed of *B. malabarica* is when yellowish-green colour of pods turned to blackish-green, dehydrated and hard. The study recommended that concentrated sulphuric acid treatment for 30 min is the best pretreatment for enhancing seed germination and reducing germination period.

Acknowledgements

We are grateful to the Kerala Forest Research Institute for scientific support. We thank Ms. Lakshmikutty VA and Mr. Suresh MK, the support staffs of Kerala Forest Seed Centre, for providing help and cooperation during the experiment.

Author details

Pazhayaveetil Kuttanpillai Chandrasekhara Pillai*, Sanal Chalil Viswanath, Thoduvayil Karunakaran Hrideek and Aviyan Hari Jiji
Kerala Forest Research Institute, Thrissur, Kerala, India

*Address all correspondence to: pkcpillai@gmail.com

References

[1] Roxburgh W. *Bauhinia*. In: Flora Indica. Calcutta, London: Carey; 1832. pp. 318-332

[2] Bandyopadhyay S, Pathak MK. Neotypification of *Bauhinia malabarica* Roxb. (Fabaceae). Candollea. 2013;**68**(2):193-195

[3] Albert S, Sharma B. Comparative foliar micromorphological studies of some *Bauhinia* (Leguminosae) species. Turkish Journal of Botany. 2013;**37**:276-281

[4] Ponomarenko SF, Pavlova ME. Seed structure in the genus *Bauhinia* L. (fam. Fabaceae Lindlkm m.) in the context of taxonomy. Biology Bulletin of the Russian Academy of Sciences. 2003;**30**(4):361-369

[5] Khan D, Zaki MJ. Size variation and surface structure of pods and seeds of *Bauhinia racemosa* Lamk. FUUAST Journal of Biology. 2016;**6**(1):73-80

[6] Anhwange BA, Ajibola VO, Oniye SJ. Nutritional potential of the seeds of *Bauhinia monandra* (Linn). Journal of Food Technology. 2005;**3**(2):204-208

[7] Jain S, Shrivastava BK. Structural determination of seed polysaccharide of *Bauhinia racemosa* by methylation studies. Oriental Journal of Chemistry. 2005;**21**(3):601-602

[8] Sinha KN, Singh M, Kumar C. Electrophoretic study of seed storage protein in five species in Bauhinia. IOSR Journal of Pharmacy and Biological Sciences. 2012;**4**(2):8-11

[9] Phansri K, Sarthima R, Thammasirirak S, Boonchalee P, Khammuang S. Antibacterial activity of *Bauhinia acuminata* L. seed protein extract with low hemolytic activity against human erythrocytes. Chiang Mai Journal of Science. 2011;**38**(2):242-251

[10] Darne P, Mehta M, Dubey P, Prabhune A. *Bauhinia* seed oil, a novel substrate for sophorolipid production. World Journal of Pharmacy and Pharmaceutical Sciences. 2014;**3**(11):792-804

[11] Kaewamatawong R, Kitajima M, Kogure N, Takayama H. Flavanols from *Bauhinia malabarica*. Journal of Natural Medicines. 2008;**62**:364-365

[12] Kittakoop P, Kirtikara K, Tanticharoen M, Thebtaranonth Y. Antimalarial preracemosols A and B, possible biogenetic precursors of racemosol from *Bauhinia malabarica* Roxb. Phytochemistry. 2000;**55**(4):349-352

[13] Thenmozhi K, Karthika K, Manian S, Paulsamy S. Studies on in vitro antioxidant potential of pod and seed parts of *Bauhinia malabarica* Roxb. Asian Journal of Biomedical and Pharmaceutical Sciences. 2014;**4**(32):48-56

[14] Vijayakumari K, Sinddhuraju P, Janardhanan K. Chemical composition and nutrient potential of the tribal pulse (*Bauhinia malabarica* Roxb.). Plant Foods for Human Nutrition. 1993;**44**:291-298

[15] Zaka S, Saleem M, Shakir N, Khan SA. Fatty acid composition of *Bauhinia variegata* and *Bauhinia malabarica* seed oils- comparison of their physio-chemical properties. Wiley InterScience. 1983;**85**(4):169-170

[16] El-Sharkawi HM, Farghali KA. Interactive effects of nitrogen, water stress and temperature in the germination of *Bauhinia* seeds. Phyton. 1987;**27**(1):139-153

[17] Asiedu JBK, Asare-Bediako E, Taah KJ, Buah JN. Effect of pre-sowing treatments on seed germination and

establishment of *Bauhinia rufescens*. International Journal of Agricultural Science. 2011;**6**(7):584-592

[18] Mwase WF, Mvula T. Effect of seed size and pretreatment methods of *Bauhinia thonningii* Schum. on germination and seedling growth. African Journal of Biotechnology. 2011;**10**(13):5143-5148

[19] ISTA. In: Kruse M, editor. Handbook on Seed Sampling. 2nd ed. Bassersdorf, Switzerland: International Seed Testing Association; 2004

[20] Bewley JD. Seed germination and dormancy. Plant Cell. 1997;**9**:1055-1066

[21] Coolbear P. Pre-sowing treatments to improve seed performance. In: Proceedings of Seed Symposium: Seed Development and Germination. Tauranga: New Zealand Society of Plant Physiologists and Agronomy Society of New Zealand; 1991, August 1991. pp. 69-76

[22] Xu Y, Cai N, He B, Zhang R, Zhao R, Zhao W, et al. Germination and early seedling growth of *Pinus densata* mast. Provenances. Journal of Forest Research. 2016;**27**(2):283-294

[23] Czabator FJ. Germination value: An index combining speed and completeness of pine seed germination. Forest Science. 1962;**8**:386-395

[24] Djavanshir K, Pourbeik H. Germination value—A new formula. Silvae Genetica. 1976;**25**:79-83

[25] Kindt R, Muasya S, Kimotho J, Waruhiu. Tree seed supplier's dictionary: Sources of seeds and microsymbionts. In: International Centre for Research in Agro-Forestry (ICRAF). Nairobi: Kenya; 1997

[26] Malla S, Shukla S, Chakravarty S. Standardized pod collection time, pod length and pre-sowing treatment of *Albizia lebbeck* at Terai zone of West Bengal, India. Journal of Forestry Research. 2012;**21**(3):338-342

[27] Usberti R, Martins L. Sulphuric acid scarification effects on *Brachiaria brizantha*, *B. humidicola* and *Panicum maximum* seed dormancy release. Revista Brasileira de Sementes. 2007;**29**(2):143-147

[28] Soliman AS, Abbas MS. Effects of sulfuric acid and hot water pre-treatments on seed germination and seedlings growth of *Cassia fistula* L. American Eurasian Journal of Agriculture and Environmental Science. 2013;**13**(1):7-15

[29] Kher MM, Nataraj M. Effect of sulfuric acid treatment on breaking of seed dormancy and germination of Indian doum palm, *Hyphaene dichotoma*, a threatened and endemic palm. Environmental and Experimental Biology. 2015;**13**:99-101

[30] Airi S, Bhatt ID, Bhatt A, Rawal RS, Dhar U. Variations in seed germination of *Hippophae salicifolia* with different presoaking treatments. Journal of Forestry Research. 2009;**20**(1):27-30

[31] FRI. Troup's the Silviculture of Indian trees. Vol. IV. Delhi: The Controller of Publications; 1983. pp. 168-170

[32] Rodrigues CR, Rodrigues BF. Enhancement of seed germination in *Macaranga peltata* for use in tropical forest restoration. Journal of Forestry Research. 2014;**25**(4):897-901

Chapter 2

The Photosynthetic Characteristics of Wild *Cymbidium faberi* in the Qinling Mountains of Central China

Junyang Song and Ning Zhang

Abstract

The large flowers of orchids make them popular as cultivated plants. Seven species of orchids in the genus *Cymbidium* (Orchidaceae) have been crossbred to create more than 220 hybrids that serve as popular cultivated ornamentals. The present study examined the daily variation in the patterns of the net photosynthetic rate and the photosynthetic response of wild *Cymbidium faberi* in the Qinling Mountains in northwestern China. The photosynthetic characteristics of this species were studied under natural conditions with a portable photosynthesis system. Double peaks were observed in the net photosynthetic rate with one around 09:00 and another around 17:00 in spring, as well as one around 11:00 and another around 15:00 in winter. Midday depression of photosynthesis was observed in wild *C. faberi* plants around 13:00 in both spring and winter. The net photosynthetic rate was strongly positively correlated with both stomatal conductance ($R = 0.913$) and the transpiration rate ($R = 0.659$) and weakly negatively correlated with the intercellular carbon dioxide concentration ($R = -0.094$). The results show that the light compensation point (LCP) and the light saturation point (LSP) of wild *C. faberi* were 25.78 and 384 μmol m^{-2} s^{-1}, respectively. The result provides reference for cultivation management especially in light management of *Cymbidium*.

Keywords: *Cymbidium faberi*, photosynthetic characteristics, Qinling mountains, light compensation point, light saturation point

1. Introduction

Cymbidium faberi (Orchidaceae) is one of the several traditional and famous orchid flowers in China. The Chinese have cultivated orchids for more than 2500 years. Most scientists currently recognize seven *Cymbidium* species in China: *C. sinensis* (Jackson ex Andr.) Willd., *C. ensifolium* (L.) Sw., *C. goeringii* (Rchb. f.), *C. faberi* Rolfe., *C. kanrau* Makino., *C. lianpan* Tang and F.T. Wang ex Y.S. Wu, and *C. longibracteatum* W.S. Wu & S.C. Chen [1]. To date, the British Royal Horticultural Society has registered 227 hybrids derived from Chinese orchids. Chinese orchids have been used as parents in the breeding of *C. faberi* because it is easy to grow, exhibits various flower colors and types, and gives off a sweet fragrance.

Wild populations of *C. faberi* are mainly distributed in the southern mountainous area of China. The Qinling Mountains support the most northern population of wild *Cymbidium* species in China, where light serves as one of the most important factors affecting its natural distribution, growth, and development. The Qinling Mountains, located at 32°40′–34°35′N and 105°30′–110°05′E, run through the central region of China and lie sandwiched between the Wei and Han rivers. This region also forms a natural and geographical boundary between northern and southern China. The mountains of the Tibetan Plateau rise to the west, while the Funiu and Dabie mountains lie to the east of the Qinling. The temperate climate north of the Qinling Mountains and the subtropical climate to the south result in a rich variety of natural plant resources in this region.

In recent years, many researchers have been interested in the photosynthetic characteristics of various plants in the Orchidaceae [2–8], whereas few studies have addressed the growth of *C. faberi* [3], especially for those plants growing in natural environments. The goal of the present study was to explore the daily photosynthetic patterns of *C. faberi* plants under natural conditions in both winter and spring. Wild *C. faberi* plants in the Qinling Mountains were examined to determine the net photosynthetic rate, photosynthetic response, and other physiological parameters. These included stomatal conductance, transpiration rate, intercellular carbon dioxide (CO_2) concentration, the light saturation point (LSP), and the light compensation point (LCP). The data presented in this study provide a foundation for cultivation management of *Cymbidium* orchid and the conservation of wild *C. faberi* in Qinling Mountains.

2. Materials and methods

2.1 The site of experiment

The experiment was established at Qianjiaping Village, Shangnan County, Shaanxi Province, China, and located in the eastern part of the Qinling Mountains at 33°20′42.7″N, and 110°41′0.14″E 816 m a. s. l. where typical populations of wild *C. faberi* occur. The plants chosen in the present study grew on a 43° southwest facing slope. *Quercus variabilis*, the dominant tree species in this area, reaches heights of about 25 m and has a canopy density of 0.4–0.5. Few shrubs grew under these trees. In this region, the mean annual, maximum, and minimum temperatures were 13.9, 41.3, and −13.1°C, respectively. The average annual rainfall was 829.8 mm with an average of 137 rainy days each year. The annual average relative humidity was 68.5% with a mean of 1973.5 hours of sunshine annually and a frostless period of 216 d; climatic data were collected between 1978 and 2008 [9].

2.2 The measurement of daily photosynthesis

The net photosynthetic rate (P_n, μmol m^{-2} s^{-1}), effective photosynthetic radiation (*PAR*, μmol Photons m^{-2} s^{-1}), stomatal conductance (G_s, mol m^{-2} s^{-1}), transpiration rate (T_r, mmol m^{-2} s^{-1}), and intercellular CO_2 concentration (C_i, μmol mol^{-1}) were measured in an open-flow gas exchange system (LI-6400, Li-Cor Inc., Lincoln, NE, USA).

The daily net photosynthetic rate was measured using blooming wild *C. faberi* plants in the Qinling Mountains in April 2016. The seasonal net photosynthetic rates were measured in January 2016 (winter) and April (spring) 2016, respectively. Sunny days were selected for all measurements that were made using one healthy leaf from each of the five randomly selected plants at the experiment site. The measuring time started at 08:00 and continued until 18:00 with hourly measurements taken. Each measurement was repeated 10 times.

2.3 The measurement of photosynthesis response

The photosynthetic responses were measured between 09:00 and 11:00 am on clear sunny days in April 2016 using the method of Gomes et al. [10]. The following criteria were employed: the CO_2 concentration in leaf chamber was 375 μmol/mol with a leaf chamber temperature of 27°C and air relative humidity of 68%. First, measurement of the light saturation was carried out for 30 min; photosynthetic active radiation (PAR) was set up to descend at 600, 500, 400, 300, 200, 100, 80, 60, 40, 20, 10, and 0 μmol m^{-2} s^{-1} during the tests using a 6400-02B LED light source that was designed to also capture data automatically. Each measurement of five samples was replicated 10 times with means used for analysis. The *net photosynthetic rate-photosynthetically active radiation* (P_n-*PAR*) curve of the photosynthetic response was calculated following Thornley's non-rectangular hyperbola [11]. The linear regression of the net photosynthetic rate to light intensity was calculated under 0–60 μmol m^{-2} s^{-1} of photosynthetically active radiation. The slope of the linear equation was the initial quantum yield. The light compensation point (LCP) and light saturation point (LSP) of wild *C. faberi* were calculated based on the curve of the photosynthetic response.

2.4 Data analysis

Data processing and drawing were conducted using Microsoft Excel 2003. The linear regression and correlation analyses were carried out using SPSS17.0.

3. Results and analysis

3.1 The daily patterns of net photosynthetic rate and photosynthetically active radiation

The net photosynthetic rate of wild *C. faberi* increased starting from 08:00 with an initial peak of 1.64 μmol m^{-2} s^{-1} at 09:00 and then decreased until 12:00 including a dramatic decrease between 11:00 and 12:00. Immediately after noon from 12:00 to 13:00, the net photosynthetic rate gradually decreased to its lowest point of 0.28 μmol m^{-2} s^{-1} at 13:00, i.e., midday depression or "noon break." After a short period of midday depression, the net photosynthetic rate gradually increased until peaking again at 17:00 at 0.69 μmol m^{-2} s^{-1} followed by a decrease to 0.30 μmol m^{-2} s^{-1} at 18:00 (blue curve in **Figure 1**).

In contrast, the effective photosynthetic radiation gradually increased from 08:00 to a maximum of 600.21 μmol m^{-2} s^{-1} at 12:00. After 12:00, it gradually decreased to 140.41 μmol m^{-2} s^{-1} at 18:00 (red curve in **Figure 1**). When the effective photosynthetic radiation reached peaked at noon, the net photosynthetic rate decreased to its second lowest rate of 0.36 μmol m^{-2} s^{-1} (**Figure 1**).

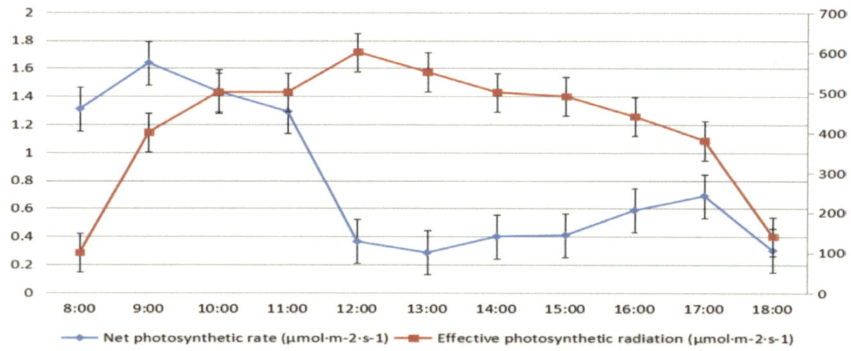

Figure 1.
The daily patterns of net photosynthetic rate and effective photosynthetic radiation in Cymbidium faberi (the error bar is standard deviation).

3.2 The relationship between net photosynthetic rate and stomatal conductance

The daily changes of stomatal conductance of wild *C. faberi* followed the same pattern as did the net photosynthetic rate (**Figure 2**). The first peak (maximum 0.0160 mol m^{-2} s^{-1}) in stomatal conductance occurred at 09:00 as did the peak in the net photosynthetic rate; however, the second peak (0.0060 mol m^{-2} s^{-1}) occurred at 16:00, 1 hour earlier than that of the net photosynthetic rate. In addition, the lowest stomatal conductance (0.0027 mol m^{-2} s^{-1}) was observed at 12:00, 1 hour earlier than that of the net photosynthetic rate (**Figure 2**). The very stomatal conductance and net photosynthetic rate were significantly correlated ($R = 0.913$ at $P < 0.01$).

3.3 The relationship between net photosynthetic and transpiration rates

The daily changes in the transpiration rate of *C. faberi* also showed a double-peak pattern with the highest peak of 0.23 mmol m^{-2} s^{-1} at 9:00 and a second peak of 0.20 mmol m^{-2} s^{-1} at 16:00 (red curve in **Figure 3**), which followed the same pattern as that of the stomatal conductance. A significant correlation was observed between the net photosynthetic and transpiration rates ($P = 0.05$; $R = 0.659$). When compared with the net photosynthetic rate, the daily changes in the transpiration rate showed a relative flat pattern.

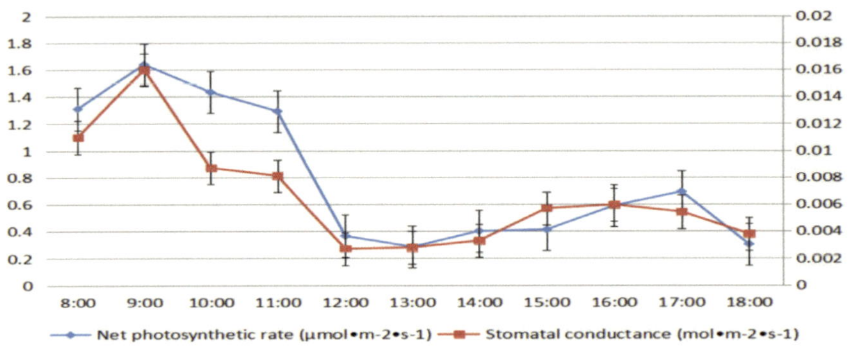

Figure 2.
The relationship between net photosynthetic rate and stomatal conductance in Cymbidium faberi during a typical day (the error bar is standard deviation).

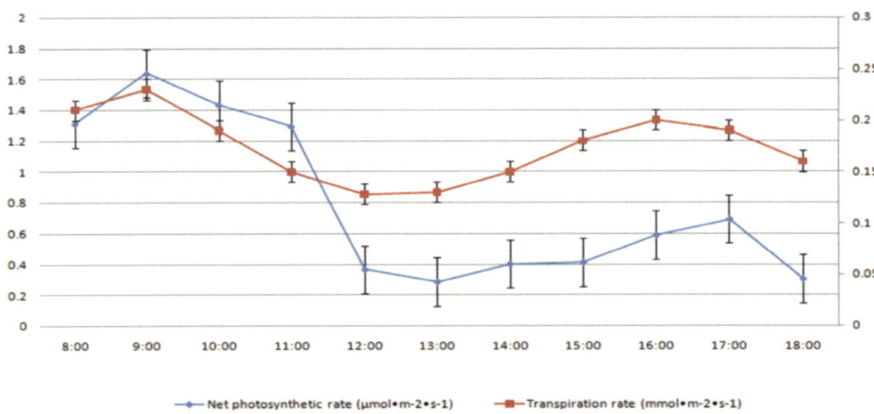

Figure 3.
The relationship between net photosynthetic rate and transpiration rate in Cymbidium faberi during a typical day (the error bar is standard deviation).

3.4 The relationship between net photosynthetic rate and intercellular CO_2 concentration

The daily changes of the intercellular CO_2 concentration of *C. faberi* (red curve in **Figure 4**) had a pattern different from that of the net photosynthetic rate. The intercellular CO_2 concentration started at 278.2 μmol mol^{-1} at 08:00 and decreased to 156.5 μmol mol^{-1} at 11:00. Around noon, the intercellular CO_2 concentration peaked at 315.1 μmol mol^{-1} and then fell dramatically to 171.8 μmol mol^{-1}. After 13:00, the intercellular CO_2 concentration gradually decreased to the lowest value of 140.3 μmol mol^{-1} at 17:00. From 17:00, the intercellular CO_2 concentration began to increase and reached 238.2 μmol mol^{-1}at 18:00.

The intercellular CO_2 concentration had an opposite pattern of change when compared with that of the net photosynthetic rate although this correlation was not significant (R = −0.094; **Figure 4**). The intercellular CO_2 concentration decreased with an increase in the net photosynthetic rate and vice versa.

3.5 Seasonal variation of net photosynthetic rate

Figure 5 shows that the patterns of the net photosynthetic rate in both spring and winter exhibited double peaks. However, the peaks of the net photosynthetic rate occurred at different times when comparing those of winter to those of spring. For example, the highest peaks in winter and spring, that is, peaks of 0.82 and 1.64 μmol m^{-2} s^{-1},

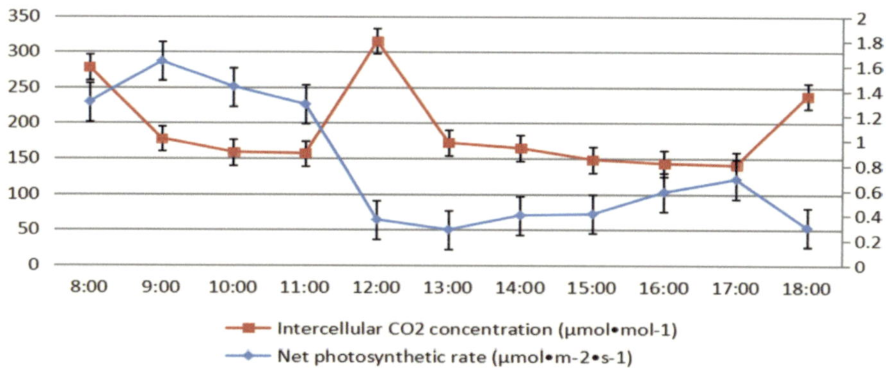

Figure 4.
The relationship between net photosynthetic rate and intercellular carbon dioxide concentration in Cymbidium faberi during a typical day (the error bar is standard deviation).

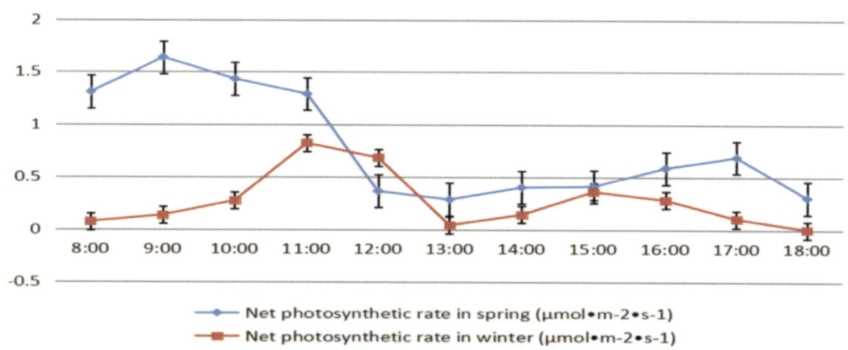

Figure 5.
Seasonal variation of net photosynthetic rate in Cymbidium faberi during a typical winter and spring day (the error bar is standard deviation).

occurred at 11:00 in winter and 09:00 in spring, respectively. The second peak of the net photosynthetic rate occurred at 15:00 and 17:00 in winter and spring, respectively. Similarly, a midday depression of photosynthesis occurred at 13:00 in both spring and winter (**Figure 5**). The average net photosynthetic rate of *C. faberi* in winter (0.26 μmol m^{-2} s^{-1}) was smaller than that in spring (0.79 μmol m^{-2} s^{-1}).

The effective photosynthetic radiation in winter and spring presented the same pattern, with an increase from 08:00 to a peak at 11:00 in winter and 12:00 in spring, and then it gradually decreased to the lowest value at 18:00 in both seasons (**Figure 6**).

3.6 Photosynthetic response curve

Figure 7 shows that the changes in the photosynthetic response P_n-PAR curves showed a parabolic shape (**Figure 7**). When the effective photosynthetic radiation ranged between 0 and 60 μmol m^{-2} s^{-1}, the change of the net photosynthetic rate presented a linear increase. However, when the effective photosynthetic radiation was between 200 and 400 μmol m^{-2} s^{-1}, the net photosynthetic rate remained at a relatively high level with a maximum of 2.41 μmol m^{-2} s^{-1} at around 400 μmol m^{-2} s^{-1}. When the effective photosynthetic radiation was greater than 400 μmol m^{-2} s^{-1}, the net photosynthetic rate obviously began to decrease, showing a light suppression phenomenon.

A quadratic equation for the photosynthetic response of *C. faberi* was obtained: $y = -0.00002x^2 + 0.01537x - 0.38298$ ($R^2 = 0.991$). Based on this equation, when

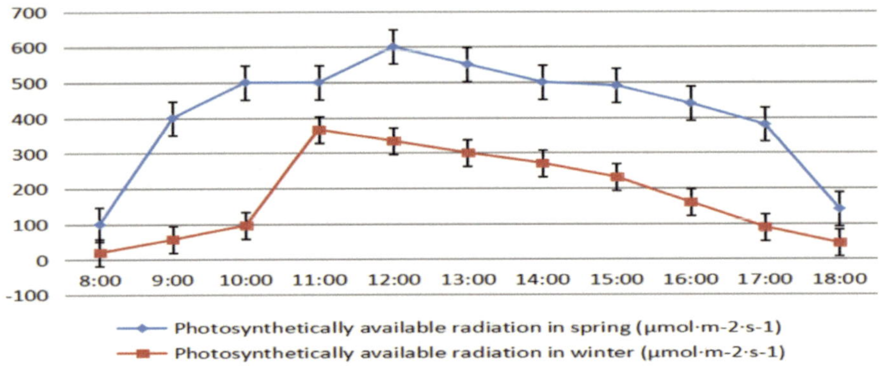

Figure 6.
Seasonal variation of effective photosynthetic radiation in Cymbidium faberi during a typical winter and spring day (the error bar is standard deviation).

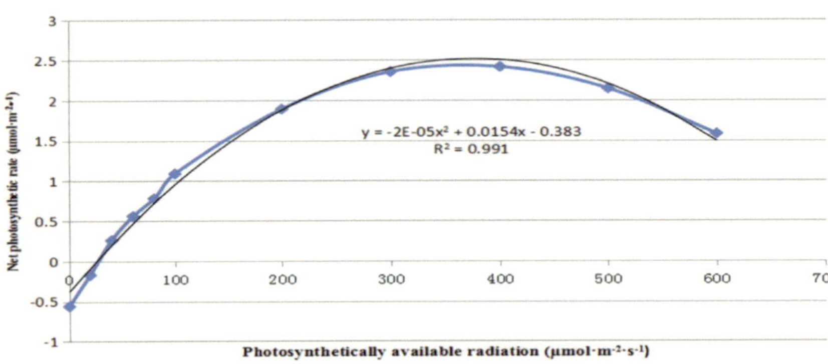

Figure 7.
Light responsive curve in Cymbidium faberi plotting photosynthetically available radiation against the net photosynthetic rate.

the effective photosynthetic radiation was 25.78 μmol m^{-2} s^{-1}, the net photosynthetic rate was zero. Also, when the effective photosynthetic radiation was 384 μmol m^{-2} s^{-1}, the net photosynthetic rate peaked (2.57 μmol m^{-2} s^{-1}). After the effective photosynthetic radiation reached to 384 μmol·m^{-2} s^{-1}, the net photosynthetic rate decreased, even when the effective photosynthetic radiation increased. Therefore, the light compensation points (LCP) and the light saturation points (LSP) of wild *C. faberi* in the Qinling Mountains were 25.78 and 384 μmol m^{-2} s^{-1}, respectively.

4. Discussion

Photosynthesis is one of the important factors for plant adaptation, substance accumulation, and metabolism. It also serves as the critical factor influencing plant growth, development, and productivity [12, 13]. However, photosynthesis is influenced by both genotype and environment as well as their interaction [14]. Multiple factors in the environment are known to interact and affect plant photosynthesis [15]. In the present study, the daily pattern of change in the net photosynthetic rate of wild *C. faberi* in the Qinling Mountains presented double peaks, as described above with a period of midday depression occurring between them. This phenomenon might be closely related to plant physiological, biochemical, and environmental factors and perhaps to other unknown factors. The midday depression in *C. faberi* might be caused by the closing of stomata in leaves at noon. In addition, the strong light intensity at noon results in the suppression of photosynthesis creating a short period of diurnal dormancy.

The factors that influence plant growth and their interactions vary at different stages of plant development [15]. Those environmental variables may cause changes in the strength of plant photosynthesis allowing plants to adapt to changes in the environment. The daily pattern and change in the net photosynthetic rate showed this rate was lower in winter than in spring in wild *C. faberi*. The peak net photosynthetic rate in spring occurred 2 hours earlier than in winter, but the second high peak was delayed by 2 hours in spring when compared with that in winter. These results are similar to those in *Carex leucochlora* [14]. The seasonal changes in the net photosynthetic rate in winter and spring were mainly caused by seasonal differences in temperature and light intensity, suggesting that the net photosynthetic rate is significantly related to environmental conditions.

The analysis of the relationship between the net photosynthetic rate and other physiological factors suggests that the net photosynthetic rate had a strong positive correction with stomatal conductance and transpiration rate and a weak negative correction with the concentration of intercellular CO_2. Hou et al. [16] and Zhang et al. [13] found that the net photosynthetic rate of *Paris polyphylla* var. *yunnanensis* was positively correlated with stomatal conductance, while Li et al. [17] found that these two were negatively correlated. Stomata provide a channel for the exchange of gasses between the cells of plant leaves and the external environment. Stomatal conductance can serve as an indicator of the degree of stomatal opening on the surface of plant leaves. Stomatal conductance and the intercellular CO_2 concentration have significant effects on plant photosynthesis and transpiration. Previous research studies have indicated that stomatal and non-stomatal restrictions can lead to a decline in the photosynthetic rate; these restrictions are differentiated by the intercellular CO_2 concentration and its pattern of change [18]. The net photosynthetic rate of the wild *C. faberi* leaf decreased with a decrease in stomatal conductance, indicating that stomatal conductance is one of the causes of this change. Stomatal conductance affects both the intercellular CO_2 concentration and the transpiration rate. Effective photosynthetic radiation and stomatal conductance, which are the main factors influencing the plant photosynthesis, combined to determine the photosynthetic rate of wild *C. faberi*.

The LCP and the LSP reflect the requirements of plants for light and light energy use. Based on the LSP and LCP values, plants with a low LCP and high LSP are adapted to a wide range of light strength, while plants with a relatively high LCP and low LSP require a narrow range of light strength [19]. This study found that the LCP and LSP of wild *C. faberi* in Qinling Mountains are 25.78 and 384 $\mu mol\ m^{-2}\ s^{-1}$, respectively, compared to the 500 and 10 $\mu mol\ m^{-2}\ s^{-1}$ in cultivated *C. faberi* species [3]. The difference of the LCP and LSP is the result of long-term adaptation to the environmental conditions under which the plants grew. Kim et al. [20] researched photosynthetic change in *Cymbidium* orchids grown under intensities of night interruption lighting. The results showed that photosynthetic photon flux of 120 $\mu mol\ m^{-2}\ s^{-1}$ was effective for *Cymbidium* orchids.

In conclusion, wild *C. faberi* plants cannot tolerate either strong or weak light, meaning it is narrowly adapted to light strength. The natural distribution of wild *C. faberi* species in Qinling Mountains is in accordance with the results of this study in that it is not found in deep shade or open sunlight. Commercial orchids were produced in greenhouse where plant growth environment can be artificially controlled. The light factor of greenhouse can be regulated and controlled by grower easily based on the range and optimum value of light factors [21]. It is beneficial to the production of orchids. Light is one of the important environmental factors that affect the growth and development of orchids. The research results of this paper show that the optimum illumination conditions for *Cymbidium* orchids are between 25.78 and 384 $\mu mol\ m^{-2}\ s^{-1}$. That is to say, light intensity should not be below 25.78 $\mu mol\ m^{-2}\ s^{-1}$ and not higher than 384 $\mu mol\ m^{-2}\ s^{-1}$.

5. Conclusions

1. The daily variation of net photosynthetic rate of wild *Cymbidium faberi* shows double-peak curve both in winter and spring.

2. There was midday depression phenomenon in the diurnal variation of photosynthetic rate for wild *Cymbidium faberi*. It appears around 13:00 pm.

3. The stomatal conductance and the net photosynthetic rate of wild *Cymbidium faberi* were significantly correlated($R = 0.913$ at $P < 0.01$).

4. There was a correlation between the net photosynthetic and transpiration rates ($R = 0.659$ at $P < 0.05$).

5. The intercellular CO_2 concentration had an opposite pattern of change when compared with that of the net photosynthetic rate although this correlation was not significant ($R = -0.094$).

6. The light compensation points (LCP) and the light saturation points (LSP) of wild *C. faberi* in the Qinling Mountains were respectively 25.78 and 384 $\mu mol\ m^{-2}\ s^{-1}$.

Acknowledgements

The authors are grateful for financial support from the State Forestry and Grassland Administration of China (Project number 2016-2046) and Forestry Department of Shaanxi Province (Project number 2013-KJ01) for this study.

Author details

Junyang Song[1*] and Ning Zhang[2]

1 College of Landscape Architecture and Arts, North West Agriculture and Forestry University, Yangling, Shaanxi, China

2 College of Water Resources and Architectural Engineering, North West Agriculture and Forestry University, Yangling, Shaanxi, China

*Address all correspondence to: songjunyang@nwsuaf.edu.cn

References

[1] Zeng S, Wu K, Zhang J, Chen Z, Duan J. Advance on research of breeding and hybridization of Chinese orchid. Chinese Wild Plant Resources. 2013;**32**:1-5

[2] Chen S, Chen Q. Research progress of photosynthetic characteristics and cultivate of *Dercdrobium officinale*. Jiangsu Agricultural Sciences. 2015;**43**:262-264

[3] Li P, Gao H, Zou Q, Wang T, Liu Y. The photosynthetic characteristics of five species of cymbidium. Acta Horticulturae Sinica. 2005;**32**:151-154

[4] Liu GH, Li K, Sun YY, Hou R, Yan H. Study of photosynthetic characteristics of *Dendrobium devonianum*. International Journal of Behavioral Development. 2014;**27**:265-269

[5] Pu X, Yin C, Zhou X, Li N, Liu Q. Changes in photosynthetic properties, ultrastructure and root vigor of *Dendrobium candidum* tissue culture seedlings during transplantation. Acta Ecologica Sinica. 2012;**32**:4114-4122

[6] Zhang L, Gao T, Zhang X, Jinping SI. Photosynthetic characteristics of *Dendrobium officinale* from five provenances. Journal of Zhejiang A & F University. 2013;**30**:359-363

[7] Zhang YY, Fang ZM, Huang WT, Zeng SJ, Kun-Lin WU, Zhang JX, et al. Study on change of leaf morphology and photosynthetic characters of three Paphiopedilum species introduced in South China. Guangdong Agricultural Sciences. 2014;**41**:92-95

[8] Zhu Q, Leng J, Qingsheng YE. Study on photosynthetic characteriestics of *Dendrobium chrysanthum* and *Dendrobium dixanthum*. Journal of South China Normal University. 2013;**45**:97-103

[9] Zhang Q, Bai H, Hua S, Zhang S. Spatial-temporal changes of forest landscape types in the eastern and western counties of Qinling mountain during the last 30 years. Huanjing Kexue Xuebao. 2010;**30**:1101-1106

[10] Gomes F, Oliva M, Mielke M, De Almeida AF, Leite H. Photosynthetic irradiance-response in leaves of dwarf coconut palm (*Cocos nucifera* L.'nana', Arecaceae): Comparison of three models. Scientia Horticulturae. 2006;**109**:101-105

[11] Thornley JHM. Dynamic model of leaf photosynthesis with acclimation to light and nitrogen. Annals of Botany. 1998;**81**:421-430

[12] Jin T, Fu B, Liiu G, Hu C, Su C, Yu L. Diurnal changes of photosynthetic characteristics of *Hippophae rhamnoides* and the relevant environment factors at different slope locations. Acta Ecologica Sinica. 2011;**31**:1783-1793

[13] Zhang SH, Liang SW, Zhong-Jun HE, Yan-Tao D, Wang Y. Comparative analysis of ten different geographical strains of *Paris polyphylla* var. *yunnanensis* based on photosynthesis. Journal of Yunnan Agricultural University. 2012;**27**:708-715

[14] Yang XJ, Ju Ying WU, Teng WJ, Yuan XH. Daily and seasonal variation of photosynthetic characteristics of *Carex leucochlora*. Pratacultural Science. 2014;**31**:102-107

[15] Zhang X, Xu D. Seasonal changes and daily courses of photosynthetic characteristics of 18-year-old Chinese fir shoots in relation to shoot ages and positions within tree crown. Scientia Silvae Sinica. 2000;**36**:19-26

[16] Hou Y, Pan YZ, Jiang BB, Yang H, Zhou X, Li JY. Diurnal variation of photosynthetic characteristics on Calla lily (*Zantedeschia hybrida*).

Hubei Agricultural Sciences. 2011;**50**:2689-2693

[17] Li Z, Zhou X, Chen J, Lei G, Du X. Photosynthetic characteristics of *Drynaria fortunei* and its relation to environmental factors. Journal of Chinese Medicinal Materials. 2012;**35**:347-351

[18] Zhang J, Zhang Z, Zha T, Chen J, Sun G, Liu C, et al. Physiological regulations of photosynthesis in a poplar plantation on a sandy soil. Acta Ecologica Sinica. 2006;**26**:1523-1532

[19] Cai SZ, Xi LI, Pan YZ, Jing LI, Ting Xing HU, Chong YE. A study on photosynthetic characteristics and growth and development of *Iris japonica* under different illumination. Acta Pharmaceutica Sinica. 2013;**22**:264-272

[20] Kim YJ, Yu DJ, Rho H, Runkle ES, Lee HJ, Kim KS. Photosynthetic changes in cymbidium orchids grown under different intensities of night interruption lighting. Scientia Horticulturae. 2015;**186**:124-128

[21] Niu G, Kozai T, Kitaya Y. Simulation of the time courses of CO_2 concentration in the culture vessel and net photosynthetic rate of cymbidium plantlets. Transactions of ASAE. 1996;**39**:1567-1573

Section 2

Genetics and Crop Improvement

Chapter 3

Summary of a Decade of South Ethiopian Coffee Improvement Activities at Awada Coffee Research Center: Fruit of the Landrace Arabica Coffee Variety Development Strategy

Mesfin Kebede Gessese

Abstract

Previously the coffee improvement strategy of Ethiopia was aimed to develop widely adaptable and stable cultivars across all coffee growing regions of the country although there is a significant ecological variation that prevails between the major coffee growing regions. Assessing the feedback from users on the performance of released coffee cultivars, the national coffee research program realized the need to initiate coffee improvement programs for each coffee growing region that possessed specific coffee quality and fetch premium price in the world market. In effect, coffee improvement program was initiated for Awada Agricultural Research Center mandated to improve south Ethiopian coffee with the financial aid of the Government of Switzerland. To date about 580 arabica coffee accessions have been collected and maintained in the center in separate sets of collection, and are under evaluation. Forty two (set I) and 16 (set II) selections are under variety trials, 12 selections are in variety verification trial, five hybrids are under variety verification trial and four high yielding cultivars that possessed the typical quality of Yirgachefe or Sidama coffee types were released to coffee growers in the region. In this paper, coffee improvement activities, such as collection and evaluation of germplasm, variety development activities and genetic studies are reviewed.

Keywords: hybrid coffee cultivars, coffee germplasm accessions, genetic diversity analysis, coffee cultivars, Hararghe coffee

1. Introduction

For more than 2 decades, Ethiopian coffee breeding program was aimed to search for improved coffee cultivars with wider adaptation to biotic and abiotic stresses and maintain stable yield across all coffee growing regions by concentrating the breeding program and source of germplasm only in the southwestern part of the country. However, this research direction has failed especially in providing cultivars that are suitable for the coffee growing areas of the Southern and eastern part of

Ethiopia due to problem of adaptation of the released coffee berry disease (CBD) resistant cultivars in these areas. In addition, these areas possess unique quality coffee types that are inherent only in the local varieties and land races of the respective locations. Hence, the national coffee research program initiated the Landrace Arabica Coffee Variety Development Strategy to establish coffee improvement programs for each coffee growing region that possesses specific coffee quality and fetch premium price in the world market [1]. To improve the yield as well as quality of south Ethiopian coffee, collection of germplasm accessions from the representative areas was undertaken. Consequently, screening of the germplasm accessions for economically important characters commenced and some promising cultivars were identified.

Coffee is the most important export crop of the south Ethiopian region with more than 46% share of the national market. It covers more than 185,000 ha of land in 50 Woredas (districts) among which 11 are high, 7 medium and 32 are low coffee producers. Garden coffee comprises 130,000 ha, semi forest 45,000 ha and forest coffee 10,000 ha. The semi forest and forest coffee production systems are pertinent to the western part of the region. In 2005 cropping season, the annual coffee production of the region was 131,000 tons out of which 100,302 tons were exported as 60% washed and 40% dry processed (SNNPR BOA and NRD, 2008; unpublished). The average yield of coffee in the region is 500 kg/ha (for local or landrace cultivars) while that of the released coffee berry disease resistant cultivars is 800 kg/ha. Though the region is highly endowed with suitable environments, the productivity of coffee per unit area remains very low as compared to the world average. In Southeast Ethiopia (East and West Hararghe zones) coffee was observed to grow as early as 850 AD. In this area coffee is grown in homesteads under intensive management systems with an estimated average holding of less than 0.2 ha of land per family. Planting spaces are very wide and the inter-row spaces are used for intercropping of various types of crops. The major coffee growing districts of Hararghe zones such as Habro, Chercher, Wobera, Garamuleta, Harar Zuria, and Gursum are known for production of best quality coffee [2–5]. On the other hand, yield is reported to be generally low in this region due mainly to the low intensity and erratic rainfall distribution pattern, and also disease problems. The low average yield of coffee in both locations was mainly attributed due to the lack of improved cultivars, shortage of improved agronomic technologies and prevalence of diseases mainly Coffee berry disease and coffee wilt disease. Moreover, physiological problems such as die back due to absence of shade trees coupled with minimum use or absence of agricultural inputs in the smallholder coffee orchards.

Awada Agricultural Research Sub-center is situated in the Tepid to cool semi-arid mid-highland agro-ecology. It is located at about 315 km south of Addis Ababa at 6°3′ N of latitude and 38° E of longitude at an altitude of about 1740 m a.s.l nearby Yirgalem town. The area has a semi-bimodal rainfall distribution characterized by double wet and dry seasons with an average precipitation of 1342 mm per annum (1988–1998) and recent data showed 1235 mm per annum (https://en.climate-data.org/africa/ethiopia/southern-nations/yirgalem-21940/). The belg (fall) starts in mid-February and extends up to mid-May (i.e., the wet season is from March to May) and the kiremt (winter = main rainy season) extends from June to September/October (i.e., the wet season is from September to October).

The sub-center is mandated to run research activities on Southern Ethiopia coffee types in general and Sidama and Gedeo coffees in particular. Therefore, major emphasis has been given to the development and release of high yielding and disease resistant coffee cultivars that maintain the standard and/or known quality of these coffee types. Four improved cultivars (Angefa, Koti, Fayate and Odicha) were released to growers and 12 promising selections are in verification trial. To date more than 580 coffee accessions have been collected and conserved at the center and most have been characterized using the IPGRI [6] coffee descriptor.

2. Arabica coffee improvement activities at Awada Research Center

2.1 Collection, characterization, and evaluation of south Ethiopian coffee germplasm accessions

This trial was laid down on-station at Awada comprising a total number of 538 accessions. The 1994, 1995 and previous Sidama coffee collections (batch I, 206 accessions and 5 checks), that are currently being evaluated for various desirable traits (cherry yield, resistance to CBD, CWD and CLR, and quality parameters) were planted in an augmented design of four blocks and 10 trees per plot (2 m × 2 m spacing between plants, hence each plot had 36 m^2) in 1997. The 1996 collections (batch II, 56 accessions and five checks) field established in 1998 was laid down in the same pattern as indicated above. The 1997 collections (batch III), which comprise 55 accessions and three checks, were planted in 1998 in RCB design of three replications and six trees per plot.

Recently new collections were added from four districts of Sidama Administrative Zone. Hundred and twenty coffee accessions collected from Dale and Aleta Wondo districts in 2005 were transplanted in the field at Awada in July 2006 in augmented design. Similarly, 100 coffee accessions collected from Bensa and Dara districts in 2006 were transplanted in the field at Awada in July 2007. Currently the seedlings are at their required stage of growth. Batch I and II were managed in multiple stems as they were stumped due to the severe drought that occurred in 2000. Hence, yield data for 2000, 2001 and 2002 could not be obtained for these two batches.

The first batch composed of 1994, 1995 and previous collection had 4 years yield data (yield data for 2000, 2001 and 2002 not available because the trees were rejuvenated to recover the damage caused by the drought occurred in 2000 main season), the 1996 collection had 5 years yield data and the 1997 collection had 7 years yield data. In the first batch of collections, average of 4 years yield data showed the top six high yielders were well above the best standard check (744) whereas, the top 10 accessions performed better than the other four checks ranging from 17.29 to 26.30 q/ha of clean coffee (**Table 1**). Similarly in the 2nd batch of collections, the top three high yielders were well above the best standard check (744) whereas, the top 10 accessions performed better than the other four checks (**Table 2**). The combined yield data showed that all the top 10 high yielders performed better than the standard checks in the 3rd batch of collections (**Table 3**). Most were free from coffee berry disease and coffee leaf rust under visual assessment score (scored from 0 to 100%; where zero is very resistant and 100% is completely susceptible); however, few accessions scored 3 to 8% infection level under field condition. The top six accessions from batch III were under variety verification trial and another 16 (those ranked from 7 to 22 based on mean yield data) were under variety trial to be evaluated in contrasting environments. Additional 15–20 promising accessions selected from the 1st and 2nd batches of collections were promoted to variety trial in 2009.

2.2 Variety and verification trials of south Ethiopian coffee selections

Three independent experiments were undergoing under this title. The first variety trial was established in two locations; Awada (mid altitude = 1745 m) and Wonago (high altitude =1850 m), respectively, in 1997 and 1999. This trial consists

Sr. No.	Coll. No.	Clean coffee yield in Qh^{-1a}					% CBD (visual)		
		1999	2003	2004	2005	Mean	2003	2004	Mean
1	85172	31.14 (28.28)	26.91 (31.07)	22.16 (15.69)	23.87 (25.50)	26.02 (25.14)	0.00	9.31	4.66
2	3883	22.21 (22.13)	12.92 (8.04)	39.55 (41.49)	9.71 (−1.84)	21.10 (17.45)	0.01	6.44	3.23
3	4083	19.79 (16.93)	19.91 (24.07)	24.91 (18.43)	16.35 (17.98)	20.24 (19.35)	0.00	0.03	0.02
4	39772	9.65 (9.57)	22.89 (18.01)	21.40 (23.34)	24.32 (12.77)	19.57 (15.92)	0.01	0.01	0.01
5	695	17.43 (19.11)	22.69 (24.25)	21.06 (20.90)	15.33 (19.67)	19.13 (20.98)	0.00	0.00	0.00
6	85171	20.89 (18.03)	19.23 (23.39)	25.97 (19.49)	6.07 (7.70)	18.04 (17.16)	2.14	10.72	6.43
7	85,170	23.84 (20.98)	12.27 (16.43)	23.78 (17.30)	11.58 (13.21)	17.87 (16.98)	0.00	0.06	0.03
8	2783	13.97 (13.89)	13.88 (9.00)	38.42 (40.36)	3.89 (−7.66)	17.54 (13.90)	0.71	2.15	1.43
9	85298	13.22 (10.36)	10.00 (14.16)	40.91 (34.43)	5.37 (7.00)	17.38 (16.49)	0.00	0.02	0.01
10	2170	14.01 (13.93)	15.49 (10.61)	29.34 (31.28)	10.60 (−0.95)	17.36 (13.72)	0.01	0.31	0.16
LSD at									
0.05		12.66	13.70	17.78	20.50				
0.01		17.75	19.21	24.93	28.75				
Standard checks									
1	74140	10.94	21.60	12.26	7.28	13.02	0.01	0.003	0.0065
2	7487	11.01	16.61	19.71	1.47	12.20	0.07	0.018	0.044
3	7440	13.09	18.96	13.61	3.94	12.40	0.03	0.007	0.0185
4	75227	9.82	21.56	20.31	7.76	14.86	0.02	0.005	0.0125
5	744	16.92	26.84	9.56	18.62	17.99	0.04	0.011	0.0255
F test		NS	NS	NS	NS				
Cv (%)		39.74	25.18	45.73	101.8				

[a]Figures in parenthesis are adjusted values.
LSD values are used for comparison between un-replicated treatments and replicated treatments. Yield data for years 2000, 2001 and 2002 not available because the coffee trees were severely attacked by drought and thrips as a result they were deformed. Rejuvenation (stumping) was performed in year 2000 that caused 3 years delay in cherry yield.

Table 1.
Mean yield and reaction to disease of the top 10 high yielding accessions for batch I.

of 42 Arabica coffee selections collected from South Ethiopia in 1970, 1977, 1981 and 1985 and two standard cultivars used as checks. Among the 42 accessions evaluated in the study, mean yield of the top 10 accessions and the standard checks over 4 years are summarized in **Tables 4** and **5**.

At Awada, combined analysis over 4 years showed that none of the top 10 accessions did better than the best check (75227) included in the study though mean yield of the top two selections 1377 and 2081 were not significantly different ($P < 0.05$) from 75227. However, the top nine accessions performed better than the other check (744). Similarly, at Wonago, combined analysis over 5 years indicated

Sr. No.	Coll. No.	Clean coffee yield in Qh^{-1a}				% CBD (visual)
		2003	2004	2005	Mean	2004
1	96/34	23.81 (23.92)	22.03 (23.95)	28.90 (24.31)	18.68 (18.04)	0.01
2	96/33	20.64 (22.26)	38.20 (36.16)	12.94 (16.25)	17.94 (18.67)	0.01
3	96/1	32.51 (32.62)	17.62 (19.54)	19.15 (14.56)	17.32 (16.68)	3.77
4	96/23	30.11 (31.73)	13.31 (11.27)	23.62 (26.93)	16.76 (17.48)	2.03
5	96/58	32.38 (29.83)	8.97 (10.74)	21.24 (21.13)	15.65 (15.43)	0.00
6	96/11	19.51 (19.62)	22.72 (24.64)	19.78 (15.19)	15.50 (14.86)	0.00
7	96/22	22.48 (24.10)	21.29 (19.26)	16. 81 (20.12)	15.15 (15.87)	0.00
8	96/21	16.35 (17.97)	32.64 (30.60)	11.53 (14.84)	15.13 (15.85)	0.00
9	96/50	10.84 (12.46)	30.04 (28.01)	19.34 (22.65)	15.06 (15.78)	0.00
10	96/41	18.20 (19.02)	27.34 (25.68)	13.01 (14.40)	14.64 (14.77)	0.02
LSD at						
0.05 0.01		6.84 9.59	14.14 19.83	10.52 14.76		
Standard checks						
1	74140	13.37	10.22	18.62	14.07	0.006
2	7487	15.29	16.59	12.77	14.88	0.003
3	7440	14.85	12.64	12.14	13.21	0.003
4	75227	9.88	6.35	6.11	7.48	0.000
5	744	19.58	17.72	13.85	17.05	0.000
F test		HS	NS	HS		
LSD at						
0.05 0.01		6.85 8.97		6.29 8.81		
Cv (%)		18.2	43.14	32.13		

[a]Figures in parenthesis are adjusted values.
NS and HS are nonsignificant and highly significant respectively at P = 0.05 and 0.01. LSD values are used for comparison between un-replicated treatments and replicated treatments. %CBD – percent coffee berry disease infection level.

Table 2.
Mean yield and reaction to disease of the top 10 high yielding accessions for batch II.

that none of the accessions out yielded the best check (744) though mean yield of the top two selections 1377 and 2081 were not significantly different (P < 0.05) from it. Nonetheless, all the top 10 accessions did better than the other standard check (75227) though only the four of the top 10 selections significantly better than 75227. Except for two selections that showed wide adaptability both in terms of yield and resistance to CBD, the rest of the accessions performed differently confirming the fact that Ethiopian coffee landraces generally show specific location adaptation [7].

The second trial, i.e., verification of Sidama coffee selections comprising 14 accessions including two standard cultivars, was established only at two locations in a RCB design of three replications, spacing of 1.5 m × 2 m and plot size of 75 and 66 trees at Konga and Korkie demonstration sites, respectively in 2004. The first cherry yield was harvested in 2007 (**Table 6**). Out of the 12 Sidama coffee selections evaluated in both locations, all the selections except one excelled the two

Sr. No.	Collection No.	Clean coffee yield in Qh^{-1}						CBD (%)				CLR (%)		
								ABT		Visual		Visual		
		2001	2002	2003	2004	2005	Mean	2003	2004	2003	2004	2001	2003	2004
1	974	9.34	18.49	24.67	40.20	11.02	20.74	0.72	3.76	0.13	0.01	3.89	17.99	4.78
2	9722	14.14	7.96	41.19	10.79	28.84	20.59	0.00	46.80	0.01	0.01	10.68	49.01	9.06
3	979	8.66	14.24	28.18	18.56	28.02	19.53	1.33	37.40	0.02	0.00	1.14	14.44	4.51
4	9718	13.02	12.08	25.41	20.36	23.41	18.86	0.00	54.20	2.23	0.00	14.11	35.77	8.17
5	971	13.32	7.93	25.46	27.34	19.56	18.72	0.00	7.09	0.01	0.01	15.44	48.89	22.39
6	9737	9.49	13.71	21.63	25.01	23.51	18.67	1.42	10.94	0.01	0.00	1.16	30.06	2.00
7	9738	5.84	10.31	23.34	32.12	17.89	17.90	3.04	12.55	0.02	0.01	1.02	16.43	12.00
8	9745	13.21	10.23	27.12	15.81	21.82	17.64	5.80	18.48	0.48	0.00	5.11	36.11	3.11
9	9714	14.39	5.10	33.62	11.37	22.77	17.45	2.67	28.10	0.02	0.00	4.67	37.17	2.68
10	975	10.31	11.83	26.52	17.16	20.35	17.24	4.18	5.60	0.04	0.00	1.02	14.66	4.78
Standard checks														
1	744	5.18	5.03	20.55	23.50	31.09	17.07	1.23	6.99	0.01	0.00	0.92	13.14	1.85
2	75227	4.68	6.12	20.60	18.13	27.71	15.45	0.00	10.33	0.05	0.00	0.12	2.97	1.04
F test		HS	HS	HS	HS	HS	HS							
LSD at														
0.05		4.85	6.75	10.02	11.04	10.61	3.98							
0.01		6.41	8.93	13.25	14.59	14.03	5.24							
Cv (%)		30.89	49.88	29.57	43.38	40.28	39.11							

HS, highly significant at P = 0.01; ABT, attached berry test, test for coffee berry disease by artificial inoculation while the young berries are still attached on the tree; CBD, coffee berry disease; CLR, coffee leaf rust.

Table 3
Mean yield and reaction to diseases of the top 10 high yielding accessions for batch III.

Sr. No.	Accessions	Clean coffee in Qh^{-1}					CBD visual (%)			CLR visual[a]
		1999	2003	2004	2005	Combined mean (03–05)	2000	2004	Mean	2003
1	1377	19.34	21.45	22.59	23.19	22.41	1.49	0.003	0.747	9.2
2	2081	21.35	16.06	38.39	12.18	22.21	4.87	0.638	2.754	4.75
3	85188	19.29	24.36	27.24	13.26	21.62	8.93	0.381	4.653	19.85
4	3677	15.91	26.68	19.24	18.39	21.43	8.04	0.628	4.332	26.85
5	2970	12.19	23.53	25.42	14.98	21.31	10.16	0.003	5.081	13.4
6	85245	10.48	22.76	20.65	19.81	21.07	7.03	0.005	3.518	3.61
7	85180	17.90	23.43	18.15	20.31	20.63	0.48	0.000	0.238	10.3
8	2181	17.48	21.44	23.01	16.84	20.43	13.85	0.388	7.120	17.65
9	3070	12.62	28.59	21.33	10.19	20.03	33.26	2.625	17.94	14.35
10	85257	16.39	22.35	15.01	20.34	19.23	0.38	0.006	0.191	22.35
Standard checks										
1	75227	6.73	24.28	26.75	26.00	25.68	0.18	0.009	0.09	0.39 (3.49)
2	744	12.67	19.18	10.25	31.61	20.35	0.24	0.003	0.12	5.11 (12.22)
F test		HS	HS	HS	HS	HS				
LSD at										
0.05		4.64	4.61	7.70	6.12	3.59				
0.01		6.13	6.09	10.16	8.08	4.73				
Cv (%)		24.22	19.28	31.71	34.33	28.49				

[a]*Figures in parenthesis are transformed value.*
HS, highly significant at P = 0.01; CBD, coffee berry disease; CLR, coffee leaf rust.

Table 4.
Mean yield and reaction to diseases of the top 10 high yielding accessions at Awada.

standard checks in 2007 cropping season. However, relatively higher yield was recorded for all the selections in Konga as compared to Korkie; and coffee berry disease infestation was relatively higher for Konga than Korkie.

The third trial was proposed to be undertaken in two sets; consequently, seedlings of 16 selections and two standard checks were transplanted at three sites (Awada, Wonago and Kumato) in August 2006. Set II was established in 2009. The seedlings were well established in the fields of the three locations.

2.3 Coffee hybrid variety development activity

Several reports have described heterosis in *Coffea arabica* with average up to 30% hybrid F_1 cultivars [8–12]. In an effort to develop high yielding, CBD resistant coffee hybrids that possess the standard quality of Sidama and Gedeo coffee in the mid and high altitudes of the south, a hybridization experiment was initiated in 1996. Through series of observations made since 1998 for yield, CBD resistance and quality of the crosses, it was possible to identify more than eight hybrids superior over the standard checks for the traits considered. Moreover, a maximum over parent heterosis of 44.6% for yield was obtained (4 years average data) for the 15 hybrids studied (**Tables** 7 and **8**). Of these 15 hybrids, 8 of them exhibited average

Sr. No.	Collection No.	Clean coffee yield in Qh^{-1}						CBD (%)			CLR visual (%)[a]
								ABT	Visual		
		2001	2002	2003	2004	2005	Combined mean	2004	2003	2004	2003
1	85259	5.36	9.98	28.29	11.11	20.98	15.14	14.07	5.07	3.56	1.39
2	85238	8.55	9.53	23.40	10.20	15.39	13.41	57.83	0.073	0.64	3.26
3	3670	6.80	10.54	24.94	12.27	11.69	13.25	40.34	3.87	3.89	4.00
4	85294	6.53	9.69	25.87	11.98	10.98	13.01	68.19	1.45	1.09	9.66
5	85232	6.50	5.05	20.42	12.69	14.38	11.81	37.73	9.20	8.89	1.86
6	1870	4.98	4.54	23.22	10.63	15.33	11.74	49.94	2.38	1.63	14.55
7	1377	6.21	7.21	18.84	12.62	12.46	11.47	31.03	0.11	2.39	2.27
8	85257	6.11	6.07	25.47	7.93	10.24	11.16	9.50	0.47	0.26	4.20
9	2077	7.17	5.96	21.27	11.95	9.30	11.13	41.02	9.91	12.76	8.15
10	85296	7.15	7.84	21.86	7.85	10.64	11.07	48.08	28.18	7.17	9.08
Standard checks											
1	744	6.24	11.09	28.82	10.34	20.64	15.43	12.30	0.12	0.01	0.81
2	75227	4.70	7.78	16.90	10.13	14.95	10.89	3.54	0.01	0.30	0.48
F test		HS	HS	HS	HS	HS	HS				
LSD at											
0.05		3.36	3.20	6.22	4.68	5.54	2.10				
0.01		4.45	3.99	8.22	6.18	7.32	2.76				
Cv (%)		45.39	41.55	27.67	38.91	40.19	37.53				

[a]Figures in parenthesis are transformed values.
HS, highly significant at P = 0.01; CBD, coffee berry disease; CLR, coffee leaf rust; ABT, attached berry test.

Table 5.
Mean yield and reaction to diseases of the top 10 high yielding accessions at Wonago.

yield of above 15 qts/ha of clean coffee, which is well above the performance of the standard checks included in the experiment [13]. Among these eight hybrids, five (744 × 1377; 7440 × 2077; 75227 × 1377; 75227 × 2077; 75227 × 1681) were proposed to be promoted to verification trial to confirm the repeatability of their performance across locations and years since consistency of repeatability of technology performance between research stations and farmers' fields may not hold universally the same, as there is high variability among farm conditions and response to improved crop management is less favorable in farmers' fields due to many conditions [14]. Variety verification experiments are designed to compare the superiority of new cultivars identified as promising by variety trials over that of the farmers' existing practices. Ethiopian indigenous arabica coffee cultivars are location specific in terms of good performance [15]. In the case of Southeastern Ethiopia coffee improvement program, hybrids are currently being evaluated for yield, resistance to major coffee diseases and quality parameters. Based on the 4 years data both high yielders and disease resistant hybrids were identified [16]. Evidence showed that there is a wide variation in environmental conditions within the southern coffee growing areas (Sidama and Gedeo Zones) that important G × E interaction might occur [16]. Therefore, the adaptability of these hybrids should be tested across locations with larger plots to verify their response to yield, major coffee diseases and other important characters.

Sr. No.	Selections	Clean coffee yield in q/ha	
		Korkie	Konga
1	85238	2.29	4.85
2	9718	4.45	4.93
3	85237	3.45	4.63
4	974	3.63	5.68
5	1377	5.22	5.75
6	9744	4.17	5.56
7	744*	1.28	4.33
8	979	3.18	8.35
9	9722	2.07	6.97
10	85294	5.61	5.04
11	85259	1.21	2.15
12	85257*	2.28	4.99
13	971	3.41	8.43
14	75227	1.30	2.58

**Selections 744 and 75227 were standard checks (released CBD resistant cultivars).*

Table 6.
Mean yield and reaction to diseases of the two selections and two standard checks.

3. The released south Ethiopian coffee cultivars

3.1 Angefa (breeder's reference: selection 1377)

Awada Agricultural Research Center (AARC) has released an improved cultivar named "Angefa" in 2006; which was high yielder and well adapted to Sidama and Gedeo coffee growing areas. This cultivar was originated from this region and represents the local coffee types plus positive advantages, i.e., resistant to coffee berry disease, high coffee bean yield (24 qt/ha on research station of Awada) and also superior qualities [17].

In relation to the previously released coffee berry disease resistant cultivars originated southwest of Ethiopia, Angefa is highly preferred by coffee farmers of Sidama and Gedeo Zones for its high vigor, yield advantage and quality characters and it fits in well with the government's strategy of strengthening the development of local landrace coffee varieties. Currently Awada Agricultural Research Center is the only source of seed for this cultivar and the demand for this cultivar in the country is very high. Angefa was initially collected from Quoti Kebele of Wonago district in Gedeo Zone of south Ethiopian region. It can be described as follows; it has an open type of growth habit, bronze leaf tip color, can grow at an altitude range of 1700–2000 m. The rainfall requirement of this cultivar is well above 1200 mm per annum. It grows best in Nitosol type of soil with the application of 125 kg DAP and 81 kg of Urea fertilizers per hectare. It can give 11 to 17 quintals of clean coffee per hectare on farmers' field. It requires 50% shade using common shade trees like Milletia, Cordia, Albizia, Sesbania and Acasia species. A spacing of 2 m × 2 m is the best recommended practice as the cultivar has open type of growth habit. In reaction to major coffee diseases, it is resistant to CBD and moderately

Sr. No.	Hybrids	Clean coffee yield in Qh^{-1a}					
		2003	2004	2005	Combined mean	% Heterosis	
						OMP	OBP
1	744 × 1681	34.78	39.44	31.04	35.08	59.85	50.17
2	1377 × 1681	29.41	45.15	20.41	31.66	36.17	35.53
3	75227 × 1681	24.77	48.08	18.19	30.34	62.12	29.88
4	744 × 2077	29.74	18.58	40.55	29.63	52.54	44.32
5	7440 × 75227	35.85	11.12	40.89	29.29	89.09	81.73
6	7440 × 1681	27.82	32.84	25.07	28.57	41.89	22.3
7	7440 × 1377	30.01	17.34	37.71	28.35	41.57	22.35
8	75227 × 1377	27.61	24.04	32.94	28.20	51.57	21.87
9	7440 × 2077	26.18	25.98	32.15	28.11	59.58	53.44
10	744 × 1377	29.75	16.81	35.58	27.38	25.40	18.32
11	75227 × 2077	22.61	34.05	23.78	26.82	65.61	46.4
12	2077 × 1681	20.66	40.85	14.69	25.40	21.88	8.73
13.	1377 × 2077	23.41	21.31	28.93	24.55	18.43	6.09
14.	744 × 7440	26.70	9.12	36.73	24.18	29.17	17.78
15	744 × 75227	25.00	14.65	26.02	21.89	26.53	6.62
Parents							
16	744	18.71	12.68	30.21	20.53		
17	7440	19.19	9.03	22.51	16.91		
18	75227	10.45	18.50	13.26	14.07		
19	1377	22.89	18.71	27.81	23.14		
20	2077	13.36	23.07	18.54	18.32		
21	1681	20.19	36.03	13.86	23.36		
F test		HS	HS	HS	HS		
LSD at							
0.05		7.77	10.19	12.47	5.85		
0.01		10.33	13.55	16.60	7.72		
CV (%)		22.2	29.25	32.01	28.45		

HS, highly significant at P = 0.01; OMP, heterosis over mid-parent; OBP, heterosis over better parent.

Table 7.
Mean yield of hybrids between south Ethiopian and southwest Ethiopian coffee genotypes at Awada.

resistant to CLR under field conditions both at Yirgalem and Wonago areas. It is also characterized by Yirga Chefe type of cup test with best raw and roast quality.

3.2 Odicha (breeder's reference: selection 974)

Odicha was released in 2010 by AARC for mid altitude (1740–1850 m) coffee growing areas of Sidama and Gedeo zones of Southern Nations and Nationalities and Peoples Region (SNNPR). It is characterized by high cherry yield potential (above the checks) and moderately resistant to coffee berry disease and highly resistant to coffee leaf rust under field conditions. It is also moderately and highly

Sr. No.	Hybrids	Clean coffee yield in Qh^{-1}									CBD visual (%)	CLR visual[a] (%)
		2000	2001	2002	2003	2004	2005	Combined mean	Heterosis (%)		2003	2003
									OMP	OBP		
1	75227 × 1377	4.76	15.28	21.63	28.82	20.79	34.81	21.02	65.38	51.77	3.81	3.81 (10.75)
2	75227 × 1681	5.65	12.47	23.45	28.54	27.74	26.46	20.72	99.81	79.08	2.21	2.21 (8.38)
3	7440 × 2077	7.52	8.32	28.21	16.78	36.61	19.27	19.45	163.91	48.13	2.52	2.52 (8.38)
4	744 × 1681	4.53	8.48	23.52	24.53	26.74	24.16	18.66	69.25	44.88	0.23	0.225 (2.5)
5	75227 × 2077	5.26	12.15	23.85	20.60	25.21	24.81	18.64	182.85	61.11	0.16	0.16 (1.11)
6	744 × 2077	6.08	11.60	24.49	15.80	27.21	25.98	18.53	155.76	43.87	1.10	1.21 (5.93)
7	7440 × 75227	6.59	8.66	26.85	22.58	22.09	21.44	18.03	45.99	37.32	0.96	0.96 (5.38)
8	744 × 1377	5.03	8.48	19.37	23.69	21.81	26.97	17.56	31.39	26.79	0.46	0.46 (3.01)
9	7440 × 1681	1.05	11.81	20.92	25.69	23.93	20.04	17.24	54.62	31.30	1.13	1.13 (4.66)
10	744 × 75227	4.68	9.11	21.08	18.43	22.94	24.56	16.80	37.42	30.43	1.71	1.71 (7.32)
11	7440 × 1377	0.85	9.02	17.77	20.36	20.55	26.47	15.84	17.42	14.37	2.13	2.13 (8.29)
12	1377 × 2077	3.64	10.18	20.42	14.69	26.39	18.29	15.60	101.81	12.64	0.34	0.335 (2.99)
13	744 × 7440	2.75	6.9	16.74	16.55	20.01	22.90	14.31	165.49	8.99	1.15	1.15 (5.54)
14	2077 × 1681	3.14	5.76	10.34	20.71	19.73	19.27	13.16	1.19	43.51	1.46	1.46 (6.75)
15	1377 × 1681	0.21	5.55	14.50	16.18	21.76	18.85	12.84	11.56	−7.29	0.86	0.86 (4.51)
Parents												
16	744	3.10	5.73	12.46	19.66	15.52	20.83	12.88			0.40	0.4 (3.58)
17	7440	3.82	8.65	17.84	14.20	19.27	15.05	13.13			1.06	1.06 (5.42)
18	75227	2.20	7.58	14.19	15.02	11.74	18.70	11.57			0.51	0.51 (3.44)
19	1377	3.34	8.44	14.26	21.13	14.67	21.24	13.85			1.72	4.72 (11.73)

Sr. No.	Hybrids	Clean coffee yield in Qh^{-1}									CBD visual (%)	CLR visual[a] (%)
		2000	2001	2002	2003	2004	2005	Combined mean	Heterosis (%)		2003	2003
									OMP	OBP		
20	2077	0.21	0.11	0.55	2.79	4.11	1.89	1.61			12.50	12.5 (20.7)
21	1681	2.54	5.99	6.70	11.67	16.15	12.00	9.17			0.66	0.66 (3.45)
F test		HS	HS	HS	HS	HS	HS	HS				
LSD at												
0.05		2.93	4.55	7.32	8.68	9.25	10.26	3.06				
0.01		3.90	6.05	9.74	11.54	12.30	13.64	4.03				
CV (%)		56.57	37.19	28.67	32.33	30.87	34.3	35.31				

Table 8.
Mean yield and reaction to diseases of crosses in set B at Wonago.

resistant to CBD and coffee wilt disease (CWD) in seedling inoculation test under controlled evaluation facility, respectively. Odicha is well adapted to the region even in marginal areas like Korke and has vigorous growth (with strong and tough stem and branches) and highly acceptable cup quality. It is medium open in growth habit which has very attractive appearance with manageable height and canopy diameter. This cultivar has an open type of growth habit with strong and stiff stem. The branches are very dense and uniformly distributed on the tree and are horizontally spreading. The leaves are long and medium wide. Both matured and young (leaf tips) leaves are green in color. Matured and ripe cherries are medium round sized and red in color.

Odicha represents both the typical quality profiles of Sidama and Yirgachefe coffee types which are spicy at Awada and floral at Konga. Besides its representation of quality profile of the locality, the cultivar has good and acceptable raw and cup quality profile under appropriate and recommended processing method. Generally, it has comparable overall quality standard with the checks and typical flavor profile representation than the standard checks.

Biennial bearing habit is the inherent characteristics of arabica coffee, which predominantly contributes to irregularity of bearing from year to year. Cultivar Odicha has revealed regular bearing habit within the acceptable range.

3.3 Fayate (breeder's reference: selection 971)

Fayate was also released in 2010 by AARC for mid altitude (1740–1850 m) coffee growing areas of Sidama and Gedeo zones of SNNPR. Fayate is highly resistant to CBD and better performed in highland areas like Wonago and Konga (Yirgachefe). In addition, selection 971 is resistant to CWD under greenhouse conditions; hence this selection can be preferably promoted to areas where CWD infestation is severe.

Fayate is characterized by high potential for cherry yield (above the checks) with highly consistent bearing habit that is unlike to most cultivars of arabica coffee. It is resistant to coffee berry disease both at visual and attached berry test evaluations. It is well adapted to the region and with highly vigorous plant that possesses strong stem, wider canopy, and strong and very long branches. It is characterized by open type of growth habit with strong and stiff stem. The branches are open and uniform across the tree. The type of branching habit is horizontal and spreading. Matured leaves are narrow, short in size and green in color. The leaf tips (shoot tips) are also green colored. It has round and medium sized cherries that become red when ripe.

Fayate represents the typical quality profile of Sidama coffee types which is characteristically known as spicy. It has good and acceptable raw and cup quality profile under appropriate and recommended processing method. The cultivar produces very acceptable export standard bean size grading. Generally, Fayate has comparable overall quality standard with the checks and typical flavor profile representation than the standard checks (744 and 75227). Fayate expressed a fairly regular bearing habit as evidenced by value of 36 over 7 years yield data at AARC, while the standard checks 75227 and 744 showed 72.17 and 63.70 at this location, respectively.

3.4 Koti (breeder's reference: selection 85257)

Cultivar Koti was also released in 2010 by AARC for mid altitude (1740–1850 m) coffee growing areas of Sidama and Gedeo zones of SNNPR. Like Fayate, Koti is also highly resistant to CBD and better performed in highland areas like Wonago and

Konga (Yirgachefe). Koti is characterized by an average potential for yield with fairly consistent bearing habit. It is highly resistant to coffee berry disease and well adapted to highland areas where coffee berry disease pressure is high. Its medium vigor (manageable height and medium canopy) can be an advantage for close spaced planting which is preferred by smallholder farmers. It also meets the standard quality (Yirgachefe and/or Sidama type) of cup test which is not achieved by the southwest released CBD resistant cultivars. It is characterized by intermediate growth habit and light bronze leaf tip color. This candidate also showed stability in yield (in contrast to specific adaptation behavior of arabica coffee varieties) as evidenced in the variety trial conducted at Awada (mid altitude) and Wonago (high altitude).

This cultivar can be expressed by thin and flexible stem along with open type of growth habit. Branches are shallowly (openly) distributed uniformly from bottom to top. The orientation of branches is horizontal and spreading. Leaf tips are bronze while matured leaves are dark green in color and broad and long in size. The matured and ripe cherries are red colored, small sized and round shaped. In addition the cherries possess persistent calyx that is a morphological marker for resistance to coffee berry disease.

4. Genetic studies

4.1 Summary of genetic studies

Since Ethiopia is the primary center of origin and genetic diversity for *C. arabica*, there is high genetic variability for yield and yield components, disease and pest resistance, and other traits. Systematic studies conducted on genetic diversity analyses of Ethiopian coffee germplasm using morphological characters confirmed the prevalence of enormous variation for economically important traits. Cluster and principal component analyses conducted on 100 Hararghe coffee germplasm accessions collected from 16 districts of East and West Hararghe Administrative Zones with four standard checks using 14 quantitative characters produced six clusters. The number of accessions per cluster ranged from five in cluster VI to 44 in cluster III. Moreover, the first four principal components explained 78.5% of the total variation prevalent within the germplasm accessions out of which 38.5% was explained by the first principal component. Similarly, a field experiment conducted on evaluation of 41 south Ethiopian coffee accessions with two standard checks of the southwest Ethiopian origin using seven morphological agronomic characters, average of 3 years data on severity of CBD and CLR infestations and clean coffee produced nine clusters. The number of accessions per cluster ranged from one in cluster IX to 13 in cluster II. Further, the first four principal components explained 82.63% of the total variation prevalent within the germplasm accessions out of which 32.52% was explained by the first principal component. The clustering pattern of the accessions revealed the prevalence of genetic diversity in the south and Southeast Ethiopian coffees for the characters considered. The study highlighted the possibility of using accessions of the distant clusters as potential candidates for the genetic improvement of both coffee types through crossing and selection. However, these reports shall be further confirmed through molecular techniques of genetic diversity analysis using the same material or germplasm.

4.2 Introduction to genetic studies

For any crop improvement program, a breeder depends on the variability present in the germplasm collections in order to advance in production, bring about stability to different biotic and abiotic stresses or effect changes in crop characteristics, and meet breeding interest [18]. In cognizant of this fact and in order to alleviate the stated production problems, a concerted efforts of coffee germplasm collection were undertaken for 3 years (1995–1997) from different coffee growing areas of Central and Southeastern part of the south Ethiopian region which resulted in the maintenance and evaluation of more than 350 coffee germplasm accessions at Awada research sub-center. Moreover, in 1998 coffee germplasm collections were conducted from different coffee growing areas of Hararghe by JARC and as a result more than 900 accessions were collected and maintained at the center.

Several workers have estimated the extent of genetic diversity present from the different sources of arabica coffee germplasm collections. For instance, a study by [19] on second progeny arabica coffee collections of Ethiopian origin indicated the prevalence of high level of variability in morphological, agronomic and biochemical characteristics. The genetic diversity analysis conducted by [20] by employing RAPD markers on cultivated and sub-spontaneous accessions of arabica coffee confirmed the narrow genetic base of commercial cultivars (3 typica and 3 bourbon types). On the other hand, they reported the existence of large genetic diversity within the sub-spontaneous material, which consisted of 11 samples representing the different coffee growing areas in Ethiopia. Further, they have suggested the prevalence of an east–west differentiation in the Ethiopian coffee germplasm. Similarly, Montagnon and Bouharmont [21] characterized 148 arabica coffee accessions for phenotype diversity under field condition. They have evaluated the accessions using 18 different morphological and agronomic traits by employing multivariate analysis and identified two main groups in the coffee accessions. According to them, accessions of group I have a more erect branching habit, narrower leaves, and were more resistant to coffee leaf rust and coffee berry disease than accessions of group II. They further opined that group I mostly contained Ethiopian arabica coffee accessions collected from west of Great Rift Valley, whereas group II contained commonly cultivated varieties throughout the world and Ethiopian accessions collected from east of Great Rift Valley in Ethiopia (south and southeast Ethiopia).

Though the reports of the above workers indicated the prevalence of lower genetic variations in the south and southeast Ethiopian coffee types, there was no systematic study conducted to quantify and verify the level of genetic diversity in both locations using large number of samples. Rather the conclusions were drawn from evaluation of few commercial cultivars originated from southeast Ethiopian coffee growing region alone. Therefore, the objective of this review paper is to report the major findings obtained from the systematic studies conducted on coffee germplasms of both South and Southeast Ethiopian coffee growing regions so that one may readily use it in the ongoing breeding program.

4.3 Research findings

Genetic diversity analysis was conducted at Wonago Agricultural Research Sub-Station on 41 south Ethiopian coffee selections collected from six Woredas of Gedeo, Sidama, and Wolayta zones along with two released coffee berry disease (CBD) resistant cultivars originated from Southwest Ethiopia [22]. Data on seven morphological agronomic characters vis-à-vis stem girth, plant height, number of primary branches, number of stem nodes, length of longest primary branches, canopy diameter and internode length of the main stem; percent disease infestation levels on CBD and coffee leaf rust (CLF) and average of 3 years clean coffee yield was obtained on the 43 genotypes. The ANOVA showed a highly significant difference among the genotypes for the seven morphological agronomic characters and yield. Southeast Ethiopian coffee population was stated to be of narrow genetic base [20, 23], however, the findings of this study indicates the presence of wide variations among Southeast Ethiopian (Sidama, Gedeo and Wolayta) landrace coffee populations located east of the Great Rift Valley. This might be attributed to the differences in the type of collections used, i.e., forest coffee versus landraces.

The cluster analysis grouped the 41 south Ethiopian coffee selections and the two southwest Ethiopian origin CBD resistant cultivars in to nine clusters suggesting the prevalence of wide phenotypic variations in the coffee populations (**Table 9**). The number of genotypes per cluster varied from one in cluster IX to 13 in cluster II. Cluster III contained selections only from Gedeo Zone (Yirgachefe & Wonago districts). On the same manner, cluster V except one selection from Wonago, was composed of selections obtained from Sidama Zone (Dale and Aleta Wondo districts). The two CBD resistant cultivars (75227 and 744) used as checks were grouped in clusters VI and VII where each cluster had three selections. The selections from Wonago district distributed in to six clusters where seven out of 16 were grouped in cluster II. Similarly the selections from Yirgachefe district distributed in to five clusters where four out of 11 were grouped in cluster III. Relatively low mean yield and higher scores of both CBD and CLR infestations characterized cluster IX that contains only one selection from Yirgachefe district (**Table 9**).

The intra and inter-cluster distance (D^2) analysis showed a highly significant ($P < 0.01$) difference among clusters. The smallest inter-cluster distance (18.6) was observed between clusters VI and VII while the highest (134.7) was between clusters V and VIII. In most of the cases, the genotypes among the clusters were significantly ($P < 0.001$) divergent from each other. Considering the intra-cluster (within cluster) distance, no significant genetic dissimilarity was detected (data not shown).

The first four principal components with eigenvalues greater than unity explained 82.63% of the total variation among the 43 genotypes for the 10 quantitative characters measured. Principal component one accounted nearly one third (32.52%) of the total variation. In light of the results obtained from the PCA, it may be possible to deduce that more than half (53%) of the variation obtained was primarily due to number of nodes, primary branches, and plant height.

Similarly, genetic diversity analysis was conducted at Awada Agricultural Research Sub-Center on 100 coffee accessions collected from 16 districts of East and West Hararghe Administrative Zones along with four released coffee berry disease

Woreda	Clusters									Total selections per Woreda
	I	II	III	IV	V	VI	VII	VIII	IX	
Yirgachefe	—	4	4	1	—	1	—	—	1	11
Wonago	3	7	2	1	1	1	1	—	—	16
Dale	—	—	—	—	2	—	—	—	—	2
Aleta Wondo	1	2	—	—	2	—	1	1	—	7
Sodozuria	—	—	—	1	—	—	—	—	—	1
Bolososore	—	—	—	—	—	—	1	1	—	2
Southwest	—	—	—	—	—	1	—	1	—	2
Unknown	1	—	1	—	—	—	—	—	—	2
Total	5	13	7	3	5	3	3	3	1	43

Source: Mesfin et al. [22].

Table 9.
Distribution of the 43 genotypes over nine clusters based on the 10 characters considered in the study.

(CBD) resistant cultivars originated from Southwest Ethiopia [24]. Data on 14 morphological agronomic characters vis-à-vis plant height, internode length of the stem, internode length of branch, number of internodes of the stem, number of internodes on the longest primary branch, total number of internodes per plant, canopy diameter, stem diameter, leaf area, number of primary branches, angle of primary branches from the main stem, number of secondary branches, length of the longest primary branch and average length of primary branches was obtained on the 104 genotypes. The ANOVA showed a highly significant difference among the genotypes for all the 14 characters considered in the study suggesting the presence of high variability among the accessions. In view of this, it may be reasonable to state that there is a good chance to improve Hararghe coffee accessions through selection and breeding.

The cluster analysis grouped the 104 coffee germplasm accessions into six clusters (**Table 10**). The size of cluster varies from five accessions in cluster V to 44 accessions in cluster III. Clusters I, II, and IV contained accessions mainly from the Western Hararghe districts whereas clusters III and V had almost equal number of accessions from both east as well as West Hararghe districts. The five accessions in cluster VI were from the two districts of West Hararghe out of which four originated from Kuni and only one from Chiro District. Three of the CBD resistant cultivars (75227, 74165 and 74140) used as checks were grouped in cluster I, where middle to high altitude accessions from Western Hararghe districts was most frequent. The fourth check, F-59 was grouped in cluster II, confirming the fact that this cultivar was distinctly different from the rest standard checks in morphology and geographical origin. It was evident that the accessions from Eastern Hararghe districts' showed close similarity (**Table 10**) with regard to their clustering patterns. For instance, the germplasm accessions from Gursum, Bedeno and Dedder Districts were found to be distributed in clusters II and III. On the other hand, accessions from Kombolcha, Girawa and Meta were scattered in clusters I, II and III where majority of their accessions were grouped in cluster III. In general, cluster III represented 58.5% of the germplasm accessions from Eastern Hararghe districts. Similarly, more than 65% of the germplasm accessions from Darolabu, Mesela and Tulo Districts of Western Hararghe were concentrated in cluster III. Accessions from Habro and Boke Districts appeared in the same clusters, i.e., clusters I, II, and III, even though, majority of their accessions appeared in the first two clusters. The germplasm accessions of Girawa, Bedeno, Kuni, Chiro, Mesela and Habro Districts were found distributed their accessions in four different clusters, which suggested that the germplasm accessions from these districts were relatively more variable. In respect to the remaining districts, the accessions were found distributed in two or three clusters, probably reflecting less variation among germplasm accessions within a particular district.

Zone	Cluster						Total accessions
	I	II	III	IV	V	VI	
East Hararge	6	6	24	2	3	—	41
West Hararge	11	13	20	7	3	5	59
South west Ethiopia*	3	1					4
Total	20	20	44	9	6	5	104

**Represented standard checks.*
Note: This table was extracted from the dendrogram. Source: Mesfin and Bayetta [24].

Table 10.
Distribution of the 104 coffee genotypes over six clusters based on quantitative traits.

Based on Mahalanobis's D^2 statistics, highly significant inter-cluster distances were obtained. Cluster II showed the maximum and significant genetic distance (102.12) from cluster VI. Further, the inter-cluster distances between clusters I and V, I and VI, II and IV, II and V, II and VI, III and VI, IV and V, and V and VI in that order were found to be highly significant. The first four principal components explained 78.5% of the total variation. Principal component one accounted more than one third (38.5%) of the variation. In light of the results obtained from principal component analysis, it may be possible to deduce that maximum variation (38.5%) accounted by principal component one was represented by such quantitative characters as length of the longest primary branch, stem diameter, total number of internodes per plant and total number of primary branches per plant.

The cluster analyses conducted for both locations, i.e., South and Southeast Ethiopia failed to clearly show relatedness of the selections due to geographical origin. Rather it is evident that there is overlapping of clustering patterns in respect of all Woredas, which could be explained as lack of differentiation among Woredas arising partly due to gene flow [25]. In light of the results obtained from the principal component analyses, it may be possible to deduce that more than half (53%) of the variation obtained in the south Ethiopian coffee was primarily due to number of stem nodes, primary branches, and plant height. Similarly, length of the longest primary branch, stem diameter, average length of primary branches, total number of internodes per plant and total number of primary branches per plant were the five important characters that contributed most to the total variation in the first principal component. This perhaps emphasized the significance of these characters to the appraisal of genetic diversity in the south and southeast Ethiopian landrace coffee populations in that order.

4.4 Conclusions and recommendations

Overlapping of the clustering patterns of the accessions from different districts of both locations indicated lack of differentiation among districts to a certain extent. Moreover, germplasm accessions from western Hararghe districts were relatively more variable in their clustering patterns as compared to eastern Hararghe districts. Based on this, it can be inferred that western Hararghe could serve as a potential source of variability for Hararghe coffee. Similarly, selections from Gedeo Zone were more divergent than selections of Sidama Zone though relatively greater number of selections was considered from Gedeo Zone. Further, it is also possible to state that quantitative characters studied significantly contributed to the elucidation of genetic diversity prevalent in the region.

The significant inter-cluster distances between clusters indicated that there is a high opportunity for obtaining transgressive segregates and maximize heterosis by crossing germplasm accessions belonging to these clusters. Therefore, the grouping of accessions by multivariate methods could be of considerable practical value to the coffee breeders so that representative accessions could be chosen from such clusters for hybridization programs. Moreover, the quantitative characters vis-à-vis number of stem nodes, primary branches, plant height, length of the longest primary branch and stem diameter could be used as a selection criterion for improving the productivity of the crop since they represent the lion's share in the variability of the coffee population in the specified area.

4.5 Gaps and challenges

The number of germplasm accessions, the locations (number of districts) and the number of characters considered for the south Ethiopian coffee were small, moreover, the germplasm collections from the Southeast Ethiopia (Hararghe zones) were appraised at pre-bearing stage only. It is however, necessary that the expression of different characters need to be studied with additional accessions over several bearing years. Furthermore, additional traits of interest and molecular techniques may be very useful in order to further confirm the present encouraging result that indicated the presence of considerable variations within South Ethiopia and Hararghe coffee populations that provides immense potential for the development of improved varieties from the local landraces for the area.

4.6 Future directions

The studies brought out that Gedeo Zone and Western Hararghe appeared to be the target areas for future intensive germplasm exploration endeavors of both locations. In the meantime, evaluation of the maintained germplasm collections for yield, quality and disease resistance must continue to provide improved cultivars for coffee growers of both regions in the shortest time possible to minimize the risk of losing smallholder coffee orchards challenged by the severe competition with chat (*Catha edulis*) especially in the Hararghe coffee growing districts.

Genetic diversity analysis using molecular techniques should be conducted on those germplasm accessions so as to confirm the results obtained and avoid duplications of accessions or genotypes. Moreover molecular markers shorten the lengthy conventional breeding scheme through the use of marker-assisted selections.

Author details

Mesfin Kebede Gessese
Department of Plant Sciences, College of Agriculture, Wolaita Sodo University, Wolaita Sodo, SNNPRS, Ethiopia

*Address all correspondence to: mesfin04@yahoo.com

References

[1] Bayetta B, Labouisse J-P. Arabica coffee (*Coffea arabica* L.) local landrace development strategy in its Center of Origin and Diversity. In: 21st Int. Sci. Colloq. On Coffee. Montpellier, France: ASIC; 2007

[2] Gure A, Chandravanshi BS, Godeto TW. Assessment of metals in roasted indigenous coffee varieties of Ethiopia. Bulletin of the Chemical Society of Ethiopia. 2018;**32**:27-38

[3] Temesgen A, Tufa A. Analysis of coffee farm productivity in Darolabu District, west Hararghe zone. Ethiopia. American Journal of Environmental and Resource Economics. 2017;**2**(4):158-161. DOI: 10.11648/j.ajere.20170204.12

[4] Sylvian PG. Some observations on *Coffea arabica* L. in Ethiopia. Turialba. 1955;**5**:37-53

[5] Brown Bridge JM, Eyassu GI. The quality of some of the main Ethiopian mild coffees. Turialba. 1968;**18**:361-372

[6] IPGRI. Descriptors for Coffee (*Coffea spp. and Psilanthus spp.*). Rome, Italy: International Plant Genetic Resources Institute; 1996

[7] Mesfin A, Bayetta B. Genotype-environment Interaction and its Implication on Selection for Improved Quality in Arabica Coffee (*Coffea arabica* L.). 17e colloque ed. Nairobi, Kenya: ASIC; 1997

[8] Srinivasan CS, Vishvashwara S. Htrosis and stability for yield in arabica coffee. Indian Journal of Genetics and Plant Breeding. 1978;**38**(3):416-420

[9] Walyaro DJA. Consideration in Breeding for Improved Yield on Quality Arabica Coffee (*Coffea arabica* L.) [PhD thesis]. Wageningen, The Netherlands: Agricultural University; 1983

[10] Mesfin A, Bayetta B. Heterosis for yield in crosses of indigenous coffee selected for yield and resistance to coffee berry disease. Acta Horticulturae. 1985;**158**:247-351

[11] Neto KA, Miguel AE, Queiroz AR. Estudo de hibridos de coffea arabica-catimor versus catuai, catindu versus catuai e outros. In: Paper presented at 19th congresso Brazilero de pesquisas cafeeiras, 23-26 de Novembro, Tres Pontas, Minas Gerais, Brazil. 1993

[12] Bertrand B, Aguilar G, Santacreo R, Anthony F, Eteinne H, Eskes AB, et al. Comportement d'hybrides F_1 de coffea arabica pour la vigueur, la production et la fetilite en Amerique centrale. In: Paper Presented at 17th Colloque International sur le café, 20–25 Juillet, Nairobi, Kenya. 1997

[13] Mohammed W. Hetrosis and Combining Ability of Yield and Yield Related Traits in Coffee (*Coffea arabica* L.) [MSc thesis]. Alemaya, Ethiopia: Alemaya University; 2004

[14] Mesfin A, Bayetta B. Genotype-environment interaction in coffee (*C. arabica* L.). In: First Ethiopian Coffee Symposium. Addis Abeba: IAR; 1986

[15] Gomez KA, Gomez AA. Statistical procedures for Agricultural Research. 2nd ed. New York: A Wiley-International Publication John Wiley and Sons, Inc.; 1984

[16] Jima Agricultural Research Center. Progress Report for the Period 2001–2003. Ethiopia: Jima; 2004

[17] Jimma Agricultural Research Center (JARC). Progress Report of Awada Research Sub-Center. Addis Ababa: EIAR; 2006

[18] IBPGR (International Board for Plant Germplasm Resources). Annual Report for 1987. Italy: Rome; 1987

[19] Catter R. Study and structure of the phenotypic variation of *Coffea arabica* from Ethiopia. In: TROPAG Data Base. 1992. p. 51

[20] Lashermes P, Trouslot P, Anthony F, Combs MC, Charier A. Genetic diversity for RAPD markers between cultivated wild accessions of Coffea arabica. Euphytica. 1996;**87**:59-64

[21] Montagnon C, Bouharmont P. Multivariatc analysis of phenotype diversity of *C. arabica* L. Genetic Resources and Crop Evolution. Vol. 43. 1996. pp. 221-227

[22] Kebede M, Bellachew B, Seifu S. Diversity in the south Ethiopian coffee. In: 21st Int. Sci. Colloq. On coffee. Montpellier, France: ASIC; 2007

[23] Anthony F, Combes MC, Herrena JC, Prakash NC, Bertrand B, Lashermes P. Genetic diversity and introgression analysis in coffee (*Coffea arabica* L.) using molecular markers. In: 16th Int. Sci. Colloq. on Coffee. Trieste, Portugal: ASIC; 2001

[24] Mesfin K, Bayetta B. Genetic divergence of Hararge coffee (*Coffea arabica* L.) Germplasm accessions at pre-bearing stage. In: 20th Int. Sci. Colloq. On coffee. Bangalore, India: ASIC; 2005

[25] Ayana A, Bekele E. Multivariate analysis of morphological variation in sorghum (*Sorghum bicolor* (L.) Moench) germplasm from Ethiopia and Eritrea. Genetic Resources and Crop Evolution. 1999;**46**:273-284

Chapter 4

Grafting in Horticultural Crop Species: Effective Pest and Disease Management Technique with Potential in Michoacan, Mexico

Juan Carlos Álvarez-Hernández

Abstract

The grafting technique is an effective alternative in crop management, specifically for the management of pests and soil pathogens; therefore, it has been recognized in all agricultural areas, which makes the a horticultural production technique more respectful with the environment. In general, this technique has been widely used in fruit growing; however, it has also been of great importance in the production of vegetables worldwide. In vegetables, the same principles applied to the grafting of fruit trees are followed, as well as specific requirements, such as controlled climatic conditions and greater care. In Michoacan, Mexico, by the phytosanitary condition in cucurbits, Solanaceae, and Caricaceae, the use of rootstocks with specific resistance characteristics offers an option for the recovery of soils, without repercussion in the environment. Although in Mexico this technique has been little exploited, in Michoacan, it is innovative in crops of Solanaceae, Cucurbitaceae, and Caricaceae. The use of grafted plants helps to improve the conditions of the crop, but also, if this technique is included in a program of integrated management of pests and diseases, it ensures the success of the production.

Keywords: approach grafting, cleft grafting, grafting, papaya, rootstock, tomato, watermelon

1. Introduction

In theory, the graft is the union of two or more pieces of living tissue, which once joined together develops as a single plant [1]. This combination of desirable characteristics consists of the removal of buds of a plant that is called graft and the root that is provided by a plant that is called rootstock [2]. The production of plant grafts has been widely expanded for fruit tree and vegetable crops, and different studies have shown that the success of the crop depends on the rootstock selected when compared with non-grafted plants [3].

In some countries, the grafting technique has been integrated into the scheme of agricultural work as an effective alternative in the management of the crops. Therefore, it has been recognized in all agricultural areas, which makes it a technique of horticultural production more respectful with the environment [4]. With this technique, the tolerance of the root system of the rootstock and the favorable productive characters of a susceptible variety are used. In vegetables the same principles applied to the grafting of fruit trees are followed, in addition to controlled environment requirements and greater post-graft care. So, the use of similar rootstocks strengthens and gives vigor to plants, therefore keeping nematodes and diseases controlled for longer than a plant that has not been grafted [5, 6].

Although there is evidence that the art of grafting was known to the Chinese from 1000 years ago before Christ [1], the grafting technique has its beginnings in the 1920s in watermelon grafted on pumpkin (*Cucurbita moschata* Duch) as rootstock, to confer resistance to wilt caused by *Fusarium*. Currently this technology is practiced successfully in Cucurbitaceae and Solanaceae in Asia and in Mediterranean countries [7]. The use of the grafting technique has been aimed at improving crops such as fruit trees and vegetables, to get their development under varied agronomic conditions [8]. It improves the resistance of crops to biotic and abiotic stresses such as salinity [9], drought tolerance [10], and nutrient deficiency [11], and this technique can be an important tool to improve fruit quality.

In Mexico, this technique is recent; however, the advantages of using it as a substitute for fumigants can counteract the main phytosanitary problems that limit the production of crops. Otherwise, in the State of Michoacan, like other states of the Mexico, the various contrasts give rise to different production systems, which favor the establishment of different crops. Despite being a predominantly agricultural territory, it has been severely affected by the production system of the monoculture type and the indiscriminate use of agrochemicals, which has caused resistance of pests and pathogens difficult to control by conventional systems. Therefore, among the management alternatives, we can see the use of the graft. Given the phytosanitary situation presented by Cucurbitaceae and Solanaceae in the State of Michoacan, the use of rootstocks with specific resistance characteristics offers an option for the recovery of soils, without repercussion in the environment. As mentioned, in our country this technique has not been fully exploited, in the State of Michoacan, it is new and innovative in the cultivations of Solanaceae, Cucurbitaceae, and Caricaceae.

2. Productive importance of tomato, watermelon, and papaya with graft potential

Mexico is located in a privileged geographic position, which favors the environmental conditions for the development of different crops in open field, and where the conditions are restrictive, crops are grown in greenhouses. Among the crops of economic importance and with potential of graft are tomato (*Solanum lycopersicum* L.), which is the second most important horticultural crop after potato; watermelon [*Citrullus lanatus* (Thunb.) Matsum. & Nakai], being an herbaceous creeping plant with 6 m in length in its branches and a highly valued product for its quality of freshness, mainly in hot seasons, and palatability in any season of the year; and papaya (*Carica papaya* L.), of the fast growing fruit species that is widespread in tropical and subtropical regions. The annual consumption *per capita* in these species is tomato, 14.3 kg; watermelon, 3.7 kg; and papaya, 6.4 kg [12].

According to SIAP-SAGARPA [13], in the last years at national level, the cultivated area has presented variable trends in tomato and watermelon, with greater amount of hectares cultivated in the year 2010, and as the years pass until the year 2017, they were reduced in 11% and 14%, respectively. However, this trend differs in the total production and the yield per hectare, since percentage of the year 2010 to the year 2017 for the tomato increased 33% and 22%, respectively, and for the

Year	Tomato			Watermelon			Papaya		
	Cultivated area (ha)	Total production (t)	Yield ($t{\cdot}ha^{-1}$)	Cultivated area (ha)	Total production (t)	Yield ($t{\cdot}ha^{-1}$)	Cultivated area (ha)	Total production (t)	Yield ($t{\cdot}ha^{-1}$)
2017	48,394	3,055,861	64.832	42,105	1,296,767	32.015	19,114	964,702	57.82
2016	48,840	2,769,611	59.336	39,903	1,129,219	30.544	19,442	957,415	56.895
2015	49,530	2,570,284	56.077	36,197	1,003,213	28.71	17,530	879,363	55.426
2014	50,850	2,320,109	48.777	35,511	955,186	28.092	16,071	840,497	57.445
2013	44,504	2,052,126	49.101	37,482	937,378	26.086	15,952	734,522	51.542
2012	55,504	2,459,874	47.102	38,288	1,011,667	27.307	16,725	680,204	49.241
2011	56,025	1,670,454	41.758	47,387	1,002,506	25.006	17,142	646,002	44.909
2010	54,238	2,058,424	42.104	48,667	1,016,215	23.375	16,261	648,235	46.49

Table 1.
The trend in tomato, watermelon, and papaya areas of cultivation, total production, and yield in Mexico country from 2010 to 2017 [13].

Year	State of Michoacan								
	Tomato			Watermelon			Papaya		
	Cultivated area (ha)	Total production (t)	Yield ($t{\cdot}ha^{-1}$)	Cultivated area (ha)	Total production (t)	Yield ($t{\cdot}ha^{-1}$)	Cultivated area (ha)	Total production (t)	Yield ($t{\cdot}ha^{-1}$)
2017	5866	211,100	36.382	990	32,337	32.680	3326	79,207	33.442
2016	6826	178,252	29.170	679	20,769	31.421	3510	70,198	32.849
2015	7845	178,931	26.204	888	21,765	24.511	2424	51,714	31.476
2014	5894	117,710	23.568	507	12,128	23.922	2128	48,605	35.094
2013	3905	73,253	24.371	676	16,500	24.427	1944	35,401	26.921
2012	5017	150,690	35.624	604	14,836	24.563	2031	43,935	33.009
2011	4768	128,367	29.013	618	15,189	24.677	2063	45,002	32.076
2010	5186	79,291	24.469	696	14,918	25.836	1998	47,947	32.999

Table 2.
The trend in tomato, watermelon, and papaya areas of cultivation, production total, and yield in the state of Michoacan, from 2010 to 2017 [13].

vvsituation for papaya, the cultivated area, the total production, and the yield per hectare, the trend always has been increasing from 2010 to 2017 (**Table 1**).

The record of the last years in the State of Michoacan has been unstable in the harvested area, total production, and yield in the three crops. However, from 2010 to 2017, the trend has been mostly upward (**Table 2**).

For its part, the various activities related to the production of tomato, watermelon, and papaya in the State of Michoacan are of great importance because they generate direct and indirect jobs, as well as being the sustenance of many families. Given the economic and social importance of these crops, their production is necessary under efficient and sustainable systems. The choice of genotypes, plantation density, phytosanitary management, and the incorporation of the grafting technique are fundamental practices to achieve higher yields and improve the quality of fruits. Nevertheless, ignorance of the correct application negatively impacts the production.

3. Current situation of potentially graft species

Tomato is one of the crops with the greatest phytosanitary problems [14], which have represented a serious problem due to the use of insecticides. This causes the death or many natural parasites of insect pests and creates genetic resistance to insecticides [15, 16]. Diseases are another limiting factor in the production of tomato [17]. Viral pathogens are disseminated by insect vectors, fungal and bacterial. Also, pathogens disseminated by seed, irrigation water and wind mean a potential danger in extensive areas of monoculture.

To achieve health in crops, measures of exclusion, eradication, and protection are used, in the context of an integrated control and use of resistant cultivars. In tomato, the theory on plant resistance has served as the basis for the development of varieties resistant to pathogens and insects, whose main source of resistance is found in wild plants [18–20]. Among the strategies to induce resistance, the conventional improvement by hybridization [21] and, another perhaps less used, the grafting technique can be distinguished [22].

Watermelon is cultivated during two cycles per year (autumn-winter and spring-summer), in irrigation and temporary. Wilt caused by *Fusarium* is considered a disease that gradually deteriorates the vigor of watermelon [23]. Also, root-knot nematodes are associated with watermelon [24]. Moreover, due to influence of agroclimatic factors and crop management, the production systems are varied [25], so the use of arbuscular mycorrhizae [26] and use of adapted genetic material, diploid or triploid hybrids [27], all contribute to obtain better yields.

Therefore, it is feasible that watermelon with management practices such as mulching, technified irrigation, and sowing methods different from the conventional one would considerably improve the productive system and competitiveness [28]. By its nature, watermelon genotypes have a high productive potential, which leads to determine their agronomic behavior to the environmental conditions of each region. Grafting technique is recognized in the agricultural ambit, and effective without negative impact on the environment, this condition is revalued with the imminent prohibition of the use of methyl bromide and its nonpolluting effect [4].

Papaya, in some stages within the production process, presents various kinds of problems. There is evidence that over time when the monocultures are continuously established, they bring with them the proliferation and resistance of pests and diseases, in which its management is difficult and has influences in the yield; also this crop requires answers oriented to the high productivity, where the densities and the nutrition play an important role. The alternatives to address the phytosanitary and physiological problems revolve around the improvement of the crop, and this can occur through the hybridization and crossings of materials, also selection of seeds, genetic engineering by including resistance genes, and in vitro propagation, all of them with the complexity of the processes and the response times. Particularly, tissue culture techniques such as micropropagation [29] both for organogenesis [30] and somatic embryogenesis have been considered for the in vitro propagation of this species; however, as biotechnological methods are until the present more expensive in relation to the use of seed, it is limited, only to hybrid genotypes that it justifies [31].

In papaya by its polygamous character, with three basic types of flowers, staminated, pistillate, and hermaphroditic, the typical propagation by seed is hindered by variability in the expression of the sexual characters and subsequent shape of the fruits. Therefore, the asexual propagation through the grafting technique would improve the papaya industry [32], since through the graft it is possible to maintain the original characteristics of the mother plants, in addition, to increase the yield, reduce height, and improve fruiting; some studies support it [33–35].

4. Grafting use in horticultural crops

In herbaceous plants, the union between the rootstock and graft is carried out by the formation of a callus of parenchymatous tissue, a structure that is differentiated to cambium tissue, which will give rise to the xylem and phloem, with which the union between vascular bundles of both individuals is restored.

It is worth mentioning that the fumigation of the soil with methyl bromide to control some soil pathogens was until recently considered one of the main factors for the success of the production of Cucurbitaceae and Solanaceae. However, the banning of methyl bromide and the lack of tolerant or resistant cultivars to biotic stress have increased the interest by the use of the grafting of vegetables [4, 36].

Some cases are mentioned on the use of grafts in the induction of resistance to pests and diseases. From the beginning, the grafting technique was used for the management of soil pathogens; currently it includes Cucurbitaceae and Solanaceae, in species of *Citrullus lanatus*, *Cucumis melo*, *Cucumis sativus*, *Solanum lycopersicum*, *Capsicum annum*, and *Solanum melongena* [37].

Cucurbitaceae are grafted on pumpkin rootstocks (*Cucurbita moschata* Duchesne and *Cucurbita maxima* Schrad) to confer resistance against soil pathogens [38]; *Phytophthora capsici* is one of the main pathogens of global importance in *C. annuum*; likewise, its management has been achieved by grafting [39], in *S. melongena* to control of *F. oxysporum* f. sp. *melongenae* [40].

On the other hand, environmental stress represents the condition with the greatest limitation for horticultural productivity and use of plants. The temperature causes economic losses of yield, also the reduction in the growth and development of the plant, caused by wilting and necrosis, affecting delay of floral induction and formation and fruit maturity [41]. According to the species of Cucurbitaceae, the threshold temperature for growth of sensitive cultures is between 8 and 12°C [42]. In the range of 25–30°C, the metabolic rates increase exponentially. Under these thresholds, many horticultural crops suffer physiological disorders which, according to intensity and length of exposure, subsequently lead to irreversible damage and death of the plant [43]. As an efficient alternative to control the temperature, the use of rootstocks is presented; since there are commercial cultivars tolerant to low temperatures, these rootstocks are recognized as a promising tool [44].

The success of the grafting depends on several factors, including union and compatibility of the graft, quality and age of seedlings, quality of the union, and post-grafting management [45]. In herbaceous species several grafting techniques have been used, and most of them coincide in some general criteria, such as performing it in the first stages of development of the plants (presence of cotyledonous leaves or first true leaves), plants under controlled conditions of temperature and environmental humidity during the period of formation of the union callus, and the subsequent acclimatization to environmental conditions.

4.1 Herbaceous grafts

The most common graft is the approach. The two individuals are sown at the same time, and when they reach 12–15 cm in height (four or five leaves), a cut is made inclined downward with a knife. This cut is made in the space of the stem between the cotyledonal leaves and the first leaves of the rootstock. The cut should be minimal and only reach half of the herbaceous stem. In the same way and in the same position, the stem of the graft is cut; instead, it will be directed upward so that the two small lips can fit as closely as possible. Finally, the grafting point is closed with small pincers or a little aluminum foil like band or some fixation device. To simplify the handling and reduce the time invested in the graft, the plants are removed from their pots before the operation and just after the graft are put back in the same pots to which soil is added, as if it were a transplant. In conditions of temperature of about 20–25°C and with a high humidity (covered with plastic bags), from 8 to 10 days, the graft will have joined, and it can proceed to cut the aerial part of the rootstock and the basal area of the graft [46].

Simple splice graft. A diagonal cut is made through the rootstock seedling just above the cotyledons. On the cut end of the pattern, a piece of thin-walled polyethylene pipe, of the appropriate diameter, is slid to give a good fit. The basal end of the scion receives a diagonal cut similar in length and inclination to that made in the pattern. The prong is slid into the plastic tube until the two cut surfaces are in close contact. The tube is held in place until healing occurs, about 12 days after grafting. If there are no leaves and the buds have not grown, the tube can be removed by sliding it over the scion; otherwise it can be cut with a razor blade [1].

In another procedure used to graft on herbaceous rootstocks, the cleft graft is used (but with a single scion), which consists of making a cut in the stem of the variety 1.5 cm below the cotyledons and making a bevel of 0.6–1.0 cm at its extreme; in the rootstock the apical bud is removed, and a slit is made between the cotyledons, to the center of the stem and down to 1–1.5 cm in length; then the graft union is inserted and tied with rubber bands or latex adhesive tape. To prevent the grafted plant from drying out, it is covered with a polyethylene bag, placed in the shade until the graft has healed, and then the plastic cover is removed [1].

Lateral slit graft. This method is practiced by making a cut in the rootstock above the first leaves and practicing a slight lateral incision directed downward (almost to the middle of the stem) along the space between the leaf that has not been cut and the two cotyledonal leaves (between 1 and 2 cm below the cutoff point). Then, the aerial part of the graft is separated, wedge-shaped and inserted into the lateral crack of the rootstock, and tied. The leaves of the rootstock are left to allow continuity of the absorption activity of the rootstock plant and to favor the union of the scion. Once the graft has been welded, it must be removed with the part of the stem that was left, as it could develop the axillary bud of the leaf and cause the graft to fail [46].

5. Grafting experiments in horticultural crops in Michoacan, Mexico

Given the phytosanitary situation presented by the horticultural species in the State of Michoacan, the use of rootstocks with specific characteristics offer an option for the recovery of soils, without repercussion in the environment.

So in integrated management, one of the strategies is plant resistance, where the technique of grafting plays fundamental importance; in Mexico, there are few documented works on grafts in vegetables and their resistance to pests and diseases [47, 48]; however, graft tests have been performed on tomato, watermelon, and papaya with spontaneous and cultivated plants and with positive results. Although it is true, in Michoacan, Mexico, this technique has not been fully exploited. In the State of Michoacan, it is new and innovative in the cultivations of Solanaceae, Cucurbitaceae, and Caricaceae.

5.1 Grafting in tomato

In Solanaceae, particularly the tomato as a species very susceptible to the attack of phytophagous insects and soil pathogens, apparently *Solanum lycopersicum var. cerasiforme* as a tomato rootstock shows resistance to the fungus *Alternaria solani* [49] and tolerance to psyllid *Bactericera cockerelli* [50] and to the aphid complex [51]. Likewise, although this rootstock comes from different geographical points, it does not demerit the phenological and fruit characteristics of the tomato placed as a graft; it also favors fruit yield [52]. These qualities served as the basis for developing the broader study [53].

Based on the above, with the objective of evaluating the incidence of the main diseases in tomato grafts on the rootstock *S. lycopersicum* var. *cerasiforme* collected in three regions of Michoacan and a collection from the State of Tabasco, Mexico, the cv. Toro® tomato commercial was used as a graft. The experiment was carried out in the Apatzingan Valley, Municipality of Paracuaro, Michoacan, Mexico. In the management of the plantation, the use of pesticides was avoided. The evaluation integrated six tomato grafts: Grafted Small Apatzingan (G-SAp), Grafted Big Apatzingan (G-BAp), Grafted Acahuato (G-Ac), Grafted Los Reyes (G-LR), Grafted Jiquilpan (G-Jiq), Grafted Tabasco (G-Tab), and Tomato (Tom)-like control. The treatments were established in field, under a completely randomized experimental block design. Weekly samplings were carried out to determine the incidence and distribution of diseases during the cycle. Sick tissue was collected to determine the causative agent. The analysis of plants with virus symptoms threw positive results, where the *Geminiviridae* group was identified. For viral diseases, incidence and severity were considered. The diseases registered were "damping off" caused by *A. solani-Fusarium* sp. complex and virus. According to the general analysis of the response of the grafts to the incidence of the present diseases, the treatment G-LR showed total resistance to "damping-off," but not to *A. solani-Fusarium* sp. complex, since to this disease, only the treatments G-SAp and G-BAp presented resistance. Regarding viral disease, all treatments were susceptible (**Table 3**). When the incidence of *Geminiviridae* was evaluated based on severity, the results were different. For example, treatments with degree of severity 3 (medium damage) were the G-BAp, G-SAp, G-Tab, and Tom, with percentages of infected plants from 5.26 to 27.27%. Level 4 (total damage) was not present (**Figure 1**).

The use of *S. lycopersicum* var. *cerasiforme* as rootstock does not influence the physical–chemical characteristics (pH, soluble solids, moisture content) of fruits in grafts compared with tomato. However, these characteristics in ecotypes of *S. lycopersicum* var. *cerasiforme* are inferior compared with the grafts and the tomato, that is, the grafted plants and the tomato have pH between 4.45 and 4.52. So, it is suggested that the pH is less acidic than that of the fruits of *S. lycopersicum* var. *cerasiforme*, which is between 4.77 and 5.37 [54]. At respect, it has been observed that in grafted tomato plants, the pH was less acidic (4.04–4.30) than in the plants without grafting (4.35–4.47) [55]; however, this variation was minimal because in

Treatment	Disease		
	Damping-off	*A. solani-Fusarium* sp. complex	Virosis
G-BAp	46.00 ± 6.92*ab	0.00 ± 0.00*a	52.67 ± 21.12**a*
Tom	25.00 ± 10.00bc	16.66 ± 10.40a	17.03 ± 13.96*a*
G-SAp	16.66 ± 11.54bc	0.00 ± 0.00a	30.33 ± 15.17*a*
G-Ac	10.23 ± 11.70cd	5.12 ± 4.43a	12.93 ± 7.88*a*
G-Jiq	8.33 ± 10.40cd	1.66 ± 2.88a	19.33 ± 7.37*a*
G-Tab	6.56 ± 6.50cd	6.66 ± 6.66a	21.00 ± 8.71*a*
G-LR	0.00 ± 0.00d	4.44 ± 3.84a	15.07 ± 13.79*a*

Means ± standard deviation, data subjected to arcsine transformation of the square root of the ratio. Different letters in the same column indicate significant differences between means (Tukey, 0.05).

Table 3.
Incidence of "damping-off," A. solani-Fusarium sp. complex and virosis in tomato grafts under field conditions. Paracuaro, Michoacan [53].

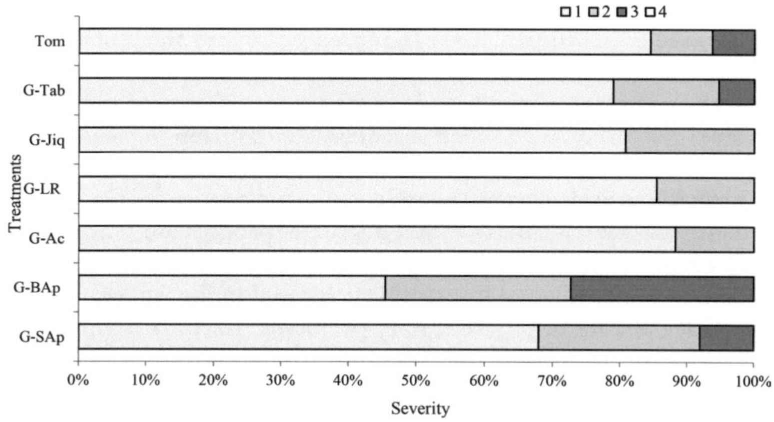

Figure 1.
Distribution of severity levels of Geminiviridae in different epidemics of tomato grafting under field conditions in Paracuaro, Michoacan: 1 = no damage, 2 = start of damage, 3 = medium damage, 4 = total damage [53].

commercial varieties, the pH is between 4.2 and 4.4 [14]; other authors [56] did not find significant statistical differences between grafts (4.33–4.41) and control 4.34.

Regarding soluble solids, the concentration in fruits was higher in *S. lycopersicum* var. *cerasiforme* with 7.5–7.75°Brix, unlike the grafts and the tomato with values 6.25–7.0°Brix, respectively [54]. The range of the cultivated varieties is between 4.5 and 5.5°Brix; although more than the varietal character, the agroecological factors influence the content of soluble solids because they can vary the °Brix for fruits of the same variety between 4 and 7 [14]. Other studies have not found differences in the °Brix of grafted and ungrafted plants [52], with values of 3.95–4.7°Brix for grafted plants and 3.95–4.95° Brix for non-grafted plants. Similarly, others report values of 3.1–4.0°Brix in grafts and 3.68°Brix in control [56]. The humidity percentage had a similar behavior only that *S. lycopersicum* var. *cerasiforme* presented lower humidity (88.39–91.33%) in comparison to the grafts and the tomato that had values of 93.99–97.44% humidity [54], differences that may be due to the wild origin of *S. lycopersicum* var. *cerasiforme*. The humidity values reported for the tomato are 94 and 95% [14, 57], which are similar to those found in our grafts.

5.2 Grafting on watermelon

The species of Cucurbitaceae that are commonly grafted are watermelon, cantaloupe, and cucumber. There are some rootstocks compatible with the three species [37]. Regarding diseases in cucurbitaceae, of the most important and that has been achieved better by grafting are those caused by pathogenic fungi, the wilt caused by *F. oxysporum* f. sp. *niveum*, being the most important. Pathogenic viruses transmitted from the soil and root-knot nematodes [58] are also important. Currently, in several countries hybrid rootstocks of wild origin or cultivated species are resistant to *Meloidogyne* spp. [4].

During the 1980s, the region of Apatzingan Valley, Michoacan, was positioned among the seven main states producing watermelon, with the advantage of presenting the ideal environment for cultivation during the autumn-winter cycle; however, their participation gradually decreased by more than 50% of the area originally intended for cultivation [59]. This reduction in agricultural land is attributed to several factors, such as the lack of more information about the evaluation and application of technical components for crop management and sustainable control of pests and diseases. Particularly, the wilt caused by *Fusarium* is considered a disease that gradually deteriorates the vigor of watermelon and cantaloupe [23]. In the Apatzingan Valley, its control came to represent 60% of the cost of cultivation, the effect of which had an impact on the quality and quantity of the crop, the reason that explains the reduction of the area dedicated to its cultivation. With respect to the root-knot nematode, it has been associated with different crops in the same region; in fact in a reported study [24], *Meloidogyne* spp. was identified in watermelon and cantaloupe and was considered a potential danger for Cucurbitaceae, since its presence causes galls in roots and decreases production. Therefore, with the objective of evaluating two rootstocks for watermelon, in two plantation distances (densities), tests were developed in a property located in the Apatzingan Valley, Michoacan, with a history of phytosanitary problems. To confirm the above in the selection of the study site in different plots, microbiological soil and root analyses were carried out to determine the existence of nematodes, bacteria, and fungi, particularly *Fusarium* (**Table 4**).

The experimental design proposed was randomized complete blocks. Six treatments were evaluated, triploid watermelon grafts on two rootstocks and triploid watermelon without grafting, all at two planting densities (4166 and 2083 plants/ha), conforming the following treatments: triploid watermelon graft on "Super Shintosa" rootstock at a density of 4167 plants per hectare (G-RSS 100), triploid

Sampled land	Nematodes			Bacteria	Presence of *Fusarium*
	Presence	Cucurbitaceae			
		Tolerance limit (No.)	Economic threshold (No.)		
Crucitas	+	2–49	≥50	2.94×10^{9} *	+
Y Griega Pozos	+			2.53×10^{7}	+
Y Griega	+			1.39×10^{6}	+
Cd. Morelos	+			3.01×10^{6}	+

**CFU/g d.s. = colony-forming units per gram of dry soil.*

Table 4.
Results of the microbiological analysis of infested soils of agricultural lands of the Apatzingan Valley, Michoacan [60].

Treatments	Triploid crunchy red					
	Soluble solids (°Brix)	Pulp hardness (kg/cm^2)	pH	Bark width (cm)	Pulp width (cm)	Moisture content (%)
G-RSS 100	11.72	1.94 bc	5.20	1.43 c	15.59 b	91.13
G-RSS 50	11.78	2.02 ab	5.30	1.45 bc	17.63 a	91.16
G-RR 100	11.74	2.12 a	5.27	1.50 ab	17.29 a	90.97
G-RR 50	11.53	2.11 a	5.27	1.45 bc	17.27 a	90.94
C-100	11.46	1.86 c	5.29	1.53 a	10.67 d	91.83
C-50	11.46	1.70 d	5.27	1.53 a	13.55 c	91.27
P	0.17	0.00	0.87	0.00	0.00	0.28
C.V.	1.59	2.79	1.91	1.42	3.93	0.51

Table 5.
Qualitative aspects of watermelon fruits grafted in two population densities [60].

watermelon graft on "Super Shintosa" rootstock at a density of 2083 plants per hectare (G-RSS 50), triploid watermelon graft on "Robusta" rootstock at a density of 4167 plants per hectare (G-RR 100), triploid watermelon graft on "Robusta" rootstock at a density of 2083 plants per hectare (G-RR 50), and triploid watermelon at a density of 4167 plants per hectare (C-100) and triploid watermelon at a density of 2083 plants per hectare (C-50) as controls. Regarding the qualitative characteristics of the fruits, the statistical analysis showed significant differences in the variables hardness of pulp, width of bark, and width of pulp, where, with the exception of the width of bark, the control treatments were exceeded in both densities. Although statistically there were differences between rootstocks (G-RSS and G-RR), with the values so close, it is presumed that the use of the graft does not alter the quality of the fruit (**Table 5**).

Regarding the phytosanitary condition, the rootstocks showed tolerance in the presence of *Fusarium* and nematodes, since in most of the variables the control was exceeded [60]. So, it is important to mention that in watermelon two main phytosanitary problems handled by the grafts are *Fusarium* wilt and nematode damage. The first case is a disease that gradually deteriorates the vigor of the plant until it is eliminated [61]. The Robusta rootstock followed by the Super Shintosa rootstock in high density favor greater efficiency related to productivity, but between the two rootstocks, Super Shintosa is sensitive to the presence of nematode [62]. It should be mentioned that watermelon is pursued to achieve the management of diseases at ground level. To avoid the use of methyl bromide, the rootstock *Cucurbita maxima* × *Cucurbita moschata* has been successfully used. Nevertheless, in the presence of nematodes, this rootstock is usually susceptible [63]. The biotic and abiotic stress of plant species derives from the soils condition and represents the greatest limitation for horticultural productivity, but when risks are minimized, it can be viable. It is important to highlight that the use of specific rootstocks to provide tolerance and/or resistance to limiting factors for the normal development of the plant is largely due to the fact that they provide a more developed and vigorous root system compared to non-grafted plants [64]. As is known, one of the main problems facing the production of watermelon in the world is the damage caused by *Fusarium*. The disinfection with methyl bromide at first gave good results, but with the time the disease generated resistance, and the use of this product has been banned; currently, the use of the graft as an alternative has reduced the problem.

5.3 Grafting in papaya

In Michoacan, ecotypes of papaya have been developed [65]. Being a predominantly agricultural territory, the region has been severely affected by the system of monoculture type and the indiscriminate use of agrochemicals, which has caused resistance of pests and diseases difficult to control through conventional systems [66]. For this reason, the Caricaceae family, particularly papaya, has the potential to be grafted to explore, in addition to the productive and phytosanitary aspect, the appearance of the sexing of plants, knowing that the preferred plants are those that emit the elongata hermaphrodite flower type and that it gives rise to elongated or marketable fruit, which is possible through grafting [67].

Therefore, in the Apatzingan Valley, experimental works of grafting in papaya were developed. The region has a semidry warm climate condition (the wettest of the semidry warm ones) with summer rains and a dominant volcanic (clayed) soil type. In order to generate and adapt a grafting method for papaya, experimental trials were carried out. Two grafting methods were tested, along with the strategies employed in vegetables, which were used for the formation of grafted papaya plants. During the development of the trials, modifications were made. In the first evaluation, two grafting methods, approach graft G.A. and cleft graft G.C. [5], and two clamping devices, lead band (G.A.B. and G.C.B.) and plastic clip (G.A.P. and G.C.P.), were compared. The response in the percentage of survival of the methods of approach and cleft grafts and fastening with lead band and clip was variable (**Figure 2**).

In the second evaluation, there were two modifications to the first graft method, called modified approach graft (G.A.M.) with two types of cuts (G.A.M.C1 and G.A.M.C2). With respect to the cleft graft, a modification was also proposed (G.C.M.) (**Figure 3**). As noted, in the second evaluation, two modifications to the evaluated methods arose, and a method called modified cleft graft (G.C.M.) was also incorporated. In **Figure 3**, the percentage values on the grafting of the graft approach methods are presented (G.A.M.C1 and G.A.M.C2). Due to its high percentage of survival (almost 90%), the treatment G.A.M.C2 is acceptable and exceeds the expectations for its use in papaya, under the conditions evaluated.

In a third evaluation, the modified approach graft method (G.A.M.C1) was tested in three containers. With respect to the election of containers, it includes the trays of 128 and 200 cavities (G.A.M.128 and G.A.M.200) and the plastic bag (G.A.M.B.). The results showed that the G.A.M.B. achieved 89% of survival (**Figure 4**).

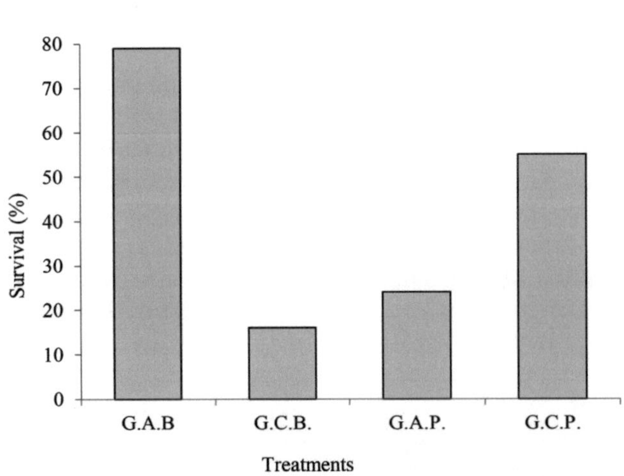

Figure 2.
Methods of approach and cleft graft and two fixation devices [68].

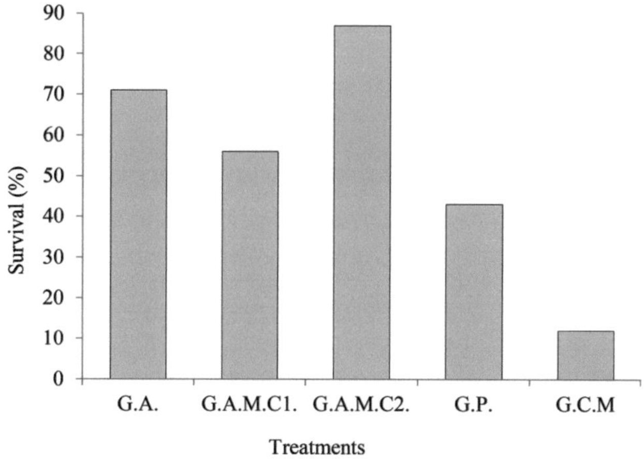

Figure 3.
Modified approach and cleft graft methods and two fixation devices [68].

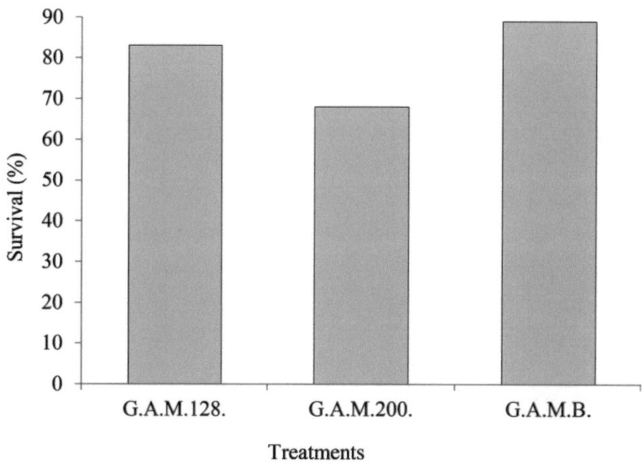

Figure 4.
Modified graft method in different containers [68].

Recapitulating, of the three evaluations, the approach graft method subjected by lead band (G.A.B.) and the modified method (G.A.M.C2.) were the most effective with 79 and 87% of survival, respectively. As for the containers used, the grafted plant with the highest yield corresponded to the use of a bag (G.A.M.B), surpassing the tray containers [68].

In another experiment, with the objective of evaluating the behavior of grafted plants and the quality of the plants, an experiment was established with five treatments formed by different combinations of rootstock and graft, in commercial genotypes. Two phases were evaluated, before and after the transplant. In the nursery, a papaya seedling was produced in a plastic bag container. The genotypes used were the varieties "Gibara" (G), "BS" (BS), "BS2" (BS2) and "Maradol" (M), and later they would serve as rootstocks (R) and grafts (G). The grafting method used was the modified approach. Five treatments were used: R. G × G. BS, R. BS × G. G, R. BS2 × G. BS, R. BS × G. BS2, and R. BS2 × G. M. The variables evaluated were the percentage of post-graft survival (prior to transplant) and percentage of post-graft survival in the field (after transplant). In the field, to determine the quality of the grafted plant in its post-graft stage, a grading scale designed under three key levels was used: N1 = vigorous, robust plant, upright leaves, normal terminal bud, does

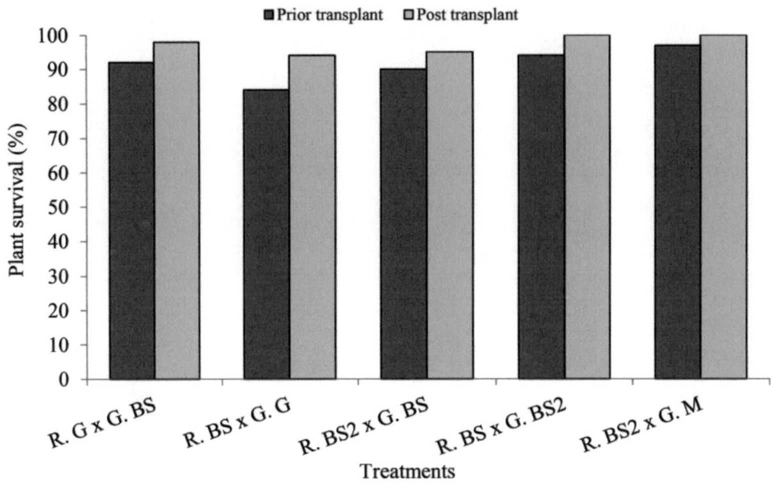

Figure 5.
Quality of grafted papaya plant prior and post transplant.

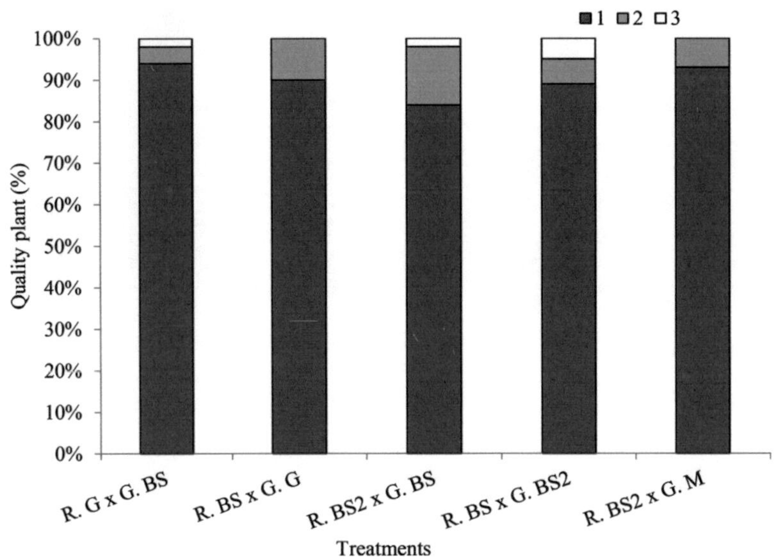

Figure 6.
Quality level of grafted papaya plant [69].

not present physical alterations in the union of the graft; N2 = vigorous plant, some leaves upright, slightly physical alterations are perceived in the union of the graft; and N3 = stressed plant appearance, weak terminal bud, contrast in stem coloration near the graft. The results obtained from the survival of the grafted papaya plant before and after the transplant are presented in **Figure 5**. The modified approach graft method responded positively in both situations, since most of the treatments exceeded well above 80% of survival, which is acceptable for the papaya species, due to its recent exploration on the subject. When making the comparison of survival between the pre- and posttransplant conditions, the values were generally lower when the pretransplant condition was registered.

In relation to plant quality, based on the three-level assessment scale, the results are presented in **Figure 6**. In general, the five grafting treatments presented level 1 (N1 = vigorous, robust plant, upright leaves, bud normal terminal, does not present physical alterations in the union of the graft) in greater percentage than levels 2 and 3;

and between treatments, R. G × G. BS, R. BS × G. G., and R. BS2 × G. M were superior with more than 90% in the first level. In level 2 (N2 = vigorous plant, some upright leaves, slightly physical alterations are perceived in the union of the graft), which was desirable to occur in a smaller proportion, only the treatments R. BS × G. G. and R. BS2 × G. BS. presented between 10 and 14%, respectively; in the rest of treatments, it was presented between 4 and 7%. This circumstance can be attributed to the fact that the plants registered under this characterization are possibly still in the postgraft recovery stage, which is caused by defect in the operation of the graft; however the situation can be reversed. Finally, level 3 (N3 = appearance of stressed plant, weak terminal bud, contrast in the coloration of the stem close to the graft), except for the treatments R. BS × G. G., and R. BS2 × G. M., did not have this condition. The other treatments presented between 2 and 5%. Although they are grafted plants that will be discarded, the percentage can be considered tolerant (**Figure 6**).

Both the modified approach graft technique and the combination of grafted genotypes in the post-graft stage before and after the transplant expressed the percentage survival condition acceptable. With the technique surpassed of the papaya graft, the bases are established to explore other aspects oriented to the management of the crop.

6. Conclusion

The species with potential for the use of graft in the Apatzingan Valley Michoacan, Mexico, are from the Solanaceae family, the tomato, the tomato from shell (*Physalis ixocarpa*), chili pepper, and the eggplant; from the Cucurbitaceae family, watermelon, cantaloupe, and cucumber; and from the Caricaceae family, the papaya, the latter in the first order to attend first to the aspect of sexing plants, where plants of the elongata hermaphrodite flower type should be selected, and in a second order to the incidence of viral diseases.

Therefore, the graft in the State of Michoacan is an alternative viable solution for the management of the mentioned crops, since it offers promising results, so its adoption can be a reality. It is also worth mentioning that the advantages of using grafted plants are much, since doing a count, is a non-polluting technique, it gives vigor to the plants, and is possible to lengthen the productive cycle. In general, the root system of the rootstocks is denser and wider; therefore, the plant has greater exploration capacity in the soil and in turn greater absorption of water and nutrients. Also, the fact of tolerating the presence of soil pests such as nematodes and harmful pathogens, plants can produce fruits and in most cases increase yields. By itself, the use of grafted plants helps to improve the conditions of the crop, but also, if this technique is included in a program of integrated management of pests and diseases, it can ensure the success of the production of different crops.

Acknowledgements

The author wished to express his gratitude to the institutions that supported and solved the development of the research works: the National Polytechnic Institute; the National Technological Institute of Mexico; the National Institute of Forestry, Agriculture and Livestock Research; as well as the team of researchers that participated directly in technical support of projects.

Author details

Juan Carlos Álvarez-Hernández
Experimental Field Apatzingan Valley, Center-Pacific Regional Research Center, National Institute of Forestry, Agriculture and Livestock Researches, Michoacan, Mexico

*Address all correspondence to: alvarez.juan@inifap.gob.mx

References

[1] Hartmann HT, Kester DE. Propagación de plantas, principios y prácticas. 2nd ed. México: CECSA; 1988. 760 p

[2] Garner RJ. The grafter's Handbook. London: Editorial Cassell; 2000. 323 p

[3] Fernández-Garcí N, Martínez V, Carvajal M. Fruit quality of grafted tomato plants grown under saline conditions. The Journal of Horticultural Science and Biotechnology. 2004;**79**(6):995-1001. DOI: 10.1080/14620316.2004.11511880

[4] González FM, Hernández A, Casanova A, Depestre T, Gómez L, Rodríguez MG. Vegetable grafting: Alternative for the management of soil pests. Revista de Protección Vegetal. 2008;**23**(2):69-74. ISSN 2224-4697

[5] Lee JM, Oda M. Grafting of herbaceous vegetable and ornamental crops. Horticultural Reviews. 2003;**28**:61-124. DOI: 10.1002/9780470650851.ch2

[6] Rivero RM, Ruiz JM, Romero L. Role of grafting in horticultural plants under stress condition. Journal of Food, Agriculture and Environment. 2003;**1**(1):70-74. DOI: 10.1234/4.2003.318

[7] Leonardi C, Romano D. Recent issues on vegetable grafting. Acta Horticulturae. 2004;(631):163-174. DOI: 10.17660/ActaHortic.2004.631.21

[8] Martínez-Ballesta MC, López-Pérez L, Hernández M, López-Berenguer C, Fernández-García N, Carvajal M. Agricultural practices for enhanced human helth. Phytochem Revviews. 2007;7(2):251-260. DOI: 10.1007/s11101-007-9071-3

[9] Fernández-García N, Martínez V, Carvajal M. Effect of salinity on growth, mineral composition, and water relations of grafted tomato plants. Journal of Plant Nutrition and Soil Science. 2004;**167**(5):612-622. DOI: 10.1002/jpln.200420416

[10] White JW, Castillo JA. Relative effect of root and shoot genotypes on yield of common bean under drought stress. Crop Science. 1989;**29**(2):360-362. DOI: 10.2135/cropsci1989.0011183X002900020026x

[11] Ruiz-Sifre G, Dole JM, Kahn B, Richardson PE, Ledford J. Correlation of poinsettia graft union development with transmission of the free-branching characteristic. Scientia Horticulturae. 1997;**69**(3-4):135-143. DOI: 10.1016/S0304-4238(97)00001-0

[12] Sagarpa editor. Food, Agriculture and Fishing Information Service (SIAP, by its acronyms in Spanish). Food and Agricultural Atlas. 1st ed. Mexico City; 2017. 231 p

[13] SIAP-SAGARPA. Statistics of National Rice Production [Internet]. 2017. Available from: http://infosiap.siap.gob.mx:8080/agricola_siap_gobmx/AvanceNacionalCultivo.do [Accessed: 2018-07-02]

[14] Nuez F. Desarrollo de nuevos cultivares. In: Nuez VF, editor. El cultivo del tomate. España: Ediciones Mundi-Prensa; 1995. pp. 626-669

[15] Liu D, Trumble JT. Tomato psyllid behavioral responses to tomato plant lines and interactions of plant lines with insecticides. Journal of Economic Entomology. 2004;**97**(3):1078-1085. DOI: 10.1603/0022-0493(2004)097[1078:TPBRTT]2.0.CO;2

[16] Casteel CL, Walling LL, Paine TD. Behavior and biology of tomato psyllid, *Bactericera cockerelli*, in response to the Mi-1.2 gene. Entomologia

Experimentalis et Applicata. 2006;**121**:67-72

[17] Tello JC, Mural J. Enfermedades no víricas del tomate. In: Nuez VF, editor. El cultivo del tomate. España: Ediciones Mundi-Prensa; 1995. pp. 524-563

[18] Hoyt E. Conserving the Wild Relatives of Crops. Delaware, United States of America: Addison-Wesley Iberoamericana; 1992. 52 p. ISBN: 0-201-51830-3

[19] Eigenbrode SD, Trumble JT. Resistance to beet armyworn, Hemipterans, and *Liriomyza* spp. in *Lycopersicon* accesiones. Journal of American society for. Horticultural Science. 1993;**118**:442-456. ISSN: 0003-1062

[20] Pérez GM, Márquez SF, Peña LA. Mejoramiento genético de hortalizas. México: Universidad Autónoma Chapingo; 1997. 380 p

[21] Restrepo SEF, Vallejo CFA, Lobo AM. Evaluación de poblaciones segregantes producidas a partir de cruzamientos entre tomate cultivado y la accesión silvestre PI134418 de *Solanum habrochaites* var. *glabratum* resistente al pasador del fruto. Acta Agronomica. 2008;**57**(1):1-8. ISSN: 2323-0118

[22] Kogan M. La resistencia de la planta en el manejo de plagas. In: Metcalf RL, Luckman WH, editors. Introducción al manejo integrado de plagas. México: Limusa-Noriega; 1990. pp. 123-172

[23] Bruton BD. Soilborne diseases in Cucurbitaceae: Pathogen virulence and host resistance. In: McCreight J, editor. Cucurbitaceae ´98. Alexandria, Virginia: ASHS Press; 1998. pp. 143-146

[24] Sosa-Moss C. *Meloidogyne incognita* nematodo agallador radicular del melón en el Valle de Apatzingán, Michoacan. Informe Técnico. Chapingo México: Colegio de Postgraduados; 1971

[25] Ban D, Zanić K, Dumicić G, Gotlin-Culjak T, Goreta-Ban S. The type of polyethylene mulch impacts vegetative growth, yield, and aphid populations in watermelon production. Journal of Food, Agriculture and Environment. 2009;7:543-550. ISSN: 1459-0263

[26] Goreta S, Perica S, Dumicić G, Bućan L, Zanić K. Growth and yield of watermelon on polyethylene mulch with different spacings and nitrogen rates. HortScience. 2005;**40**(2):366-369 ISSN: 2327-9834

[27] Canales CR, Sánchez BJA. Cadena agroalimentaria de sandía. Caracterización de los eslabones de la cadena e identificación de los problemas y demandas tecnológicas. Campeche: INIFAP. COFUPRO. FUPROCAM; 2003. p. 103

[28] Pérez-González JL, Sánchez-Cohen I, Mendoza-Moreno FS, Inzunza-Ibarra MA, Cueto-Wong JA. Productividad y rendimiento de sandía por efecto del agua en diferentes condiciones de manejo. Revista Chapingo Serie Horticultura. 2003;**9**(2):209-223. DOI: 10.5154/r.rchsh.2001.08.049

[29] Fitch MM, Leong T, Akashi L, Yeh A, White S, De la Cruz A, et al. Growth and yield of clonally propagated and seedling-derived papayas. I. Growth. II. Yield. The Journal of Horticultural Science and Biotechnology. 2005;**40**(5):1283-1290. ISSN: 2380-4084

[30] Hossain M, Rahman N, Islam S, Joarder O. High efficien plant regeneration from petiole explants of *Carica papaya* L. through organogenesis. Plant Cell Reports. 1993;**13**(2):99-102. ISSN: 1432-203X

[31] Elder RJ, Macleod WNB, Reid DJ, Gillespie RL. Growth, yield and phenology of 2 hybrid papayas (*Carica papaya* L.) as influenced by method of water application. Australian

Journal of Experimental Agriculture. 2000;**40**(5):739-746. DOI: 10.1071/EA99107

[32] Senthilkumar S, Kumar N, Soorianathasundaram K, Jeya-Kumar P. Aspects on asexual propagation in papaya (*Carica papaya* L.)- a review. Agricultural Reviews. 2014;**35**(4):307-313. DOI: 10.5958/0976-0741.2014.00919.2

[33] Allan P. Phenology and production of *Carica papaya* 'honey gold' under cool subtropical conditions. Acta Horticulturae. 2007;(740):217-223. DOI: 10.17660/ActaHortic.2007.740.26

[34] Lima de LA, Naves RV, Yamanishi OK, Pancoti HL. Behavior of three papaya genotypes propagated by grafting in Brazil. Acta Horticulturae. 2010;(851):343-348. DOI: 10.17660/ActaHortic.2010.851.52

[35] Lange AH. Reciprocal grafting of normal and dwarf solo papaya on growth and yield. Horticultural Science. 1969;**4**(4):304-306

[36] King SR, Davis AR, Liu W, Levi A. Grafting for disease resistance. HortScience. 2008;**43**(6):1673-1676 ISSN: 2327-9834

[37] King SR, Davis AR, Zhang X, Crosby K. Genetics, breeding and selection of rootstocks for solanaceae and cucurbitaceae. Scientia Horticulturae. 2010;**127**(8):106-111. DOI: 10.1016/j.scienta.2010.08.001

[38] Santos LM, Fontanetti VM, Minami K, Tessarioli NJ. Evaluation of graft supports for Japanese cucumber. Scientia Agricola. 2000;**57**(1):169-172. DOI: 10.1590/S0103-90162000000100027

[39] Santos HS, Goto R. Sweet pepper grafting to control *Phytophthora* blight under protected cultivation. Horticultura Brasileira. 2004;**22**(1):45-49. DOI: 10.1590/S0102-05362004000100009

[40] Toppino L, Vale G, Rotino GL. Inheritance of *Fusarium* wilt resistance introgresed from *Solanum aethiopicum* Gilo and Aculeatum groups into cultivated eggplant (*S. melongena*) and development of associated PCR-based markers. Molecular Breeding. 2008;**22**(2):237-250. DOI: 10.1007/s11032-008-9170-x

[41] Ahn SJ, Im YJ, Chung GC, Cho BH, Suh SR. Physiological responses of grafted-cucumber leaves and rootstock roots affected by low root temperature. Scientia Horticulturae. 1999;**81**(4):397-408. DOI: 10.1016/S0304-4238(99)00042-4

[42] Hansen LD, Afzal M, Breidenbach RW, Criddle RS. High and low temperature limits to growth of tomato cells. Planta. 1994;**195**(1):1-9. ISSN: 14322048

[43] Venema JH, Linger P, Van Heusden AW, Van Hasselt PR, Brüggemann W. The inheritance of chilling tolerance in tomato (*Lycopersicon* spp.). Plant Biology. 2005;**7**(2):118-130. DOI: 10.1055/s-2005-837495

[44] Rouphael Y, Schwarz D, Krumbein A, Colla G. Impact of grafting on product quality of fruit vegetable crops. Scientia Horticulturae. 2010;**127**(2):172-179. DOI: 10.1016/j.scienta.2010.09.001

[45] Kawaguchi M, Taji A, Backhouse D, Oda M. Anatomy and physiology of graft incompatibility in solanaceous plants. The Journal of Horticultural Science and Biotechnology. 2008;**83**(5):581-588. DOI: 10.1080/14620316.2008.11512427

[46] Peil RM. Grafting of vegetable crops. Ciencia Rural. 2003;**33**(6):1167-1169. DOI: 10.1590/S0103-84782003000600028

[47] Alvarez HJC, Cortez MH, Garcia RI. Injertos de jitomate (*Solanum lycopersicum* L.) en parientes silvestres como fuente de resistencia a plagas. In: Estrada VEG, Equihua MA, Chaires GM, Acuña SJA, Padilla RJR, Mendoza EA, editors. Entomologia Mexicana. 1ra ed. Vol. 8. Mexico: Sociedad Mexicana de Entomologia; 2009. pp. 551-555. ISBN 968-839-559-2

[48] Cortez-Madrigal H. Injertos en parientes silvestres de jitomate y su potencial en el manejo de plagas. Entomología Mexicana. 2008;7:570-574

[49] Méndez-Inocencio C, Cortez-Madrigal H, Muñoz-Ruíz C, Ochoa-Gaytan E. El tinguaraque *Lycopersicon esculentum* var. *cerasiforme* como fuente de resistencia a plagas y enfermedades en Michoacan. In: Memorias del 2do. Zamora, Michoacan: Congreso Estatal de Ciencia y Tecnología; 2006. p. 64

[50] Cortes MH, Alvarez HJC. Incidencia y daño de insectos fitófagos en injertos de jitomate sobre parientes silvestres. In: Memorias del IX Simposio Internacional y IV Congreso Nacional de Agricultura Sostenible; Boca del Rio Veracruz México; 2007. p. 296

[51] Alvarez HJC, Cortez MH. Incidencia de insectos fitófagos en injertos de jitomate (*Solanum lycopersicum* L.) sobre parientes silvestres. In: Memorias del 5to. Morelia Michoacan Mexico: Congreso Estatal de Ciencia y Tecnología; 2009. pp. 1-6

[52] Alvarez HJC, Cortez MH, Garcia RI. Injertos de jitomate en ecotipos silvestres de *Solanum lycopersicum* var. *Cerasiforme*. In: Memorias del XI. Chapingo Mexico: Congreso Nacional de Ciencias Agronómicas; 2009. pp. 1-2

[53] Alvarez HJC, Cortez MH, Ceja TLF. Incidence of diseases in grafters of tomato (*Solanum lycopersicum*) on its wild relative. In: Memorias del XI Congreso Internacional/ XXXVI. Acapulco Gerrero Mexico: Congreso Nacional de la Sociedad Mexicana de Fitopatología; 2009. pp. 1-6

[54] Alvarez HJC. Caracterización agronómica e incidencia de enfermedades de injertos de jitomate en jitomate silvestre. In: Memorias del XIII. Chapingo México: Congreso Nacional de Ciencias Agronomicas; 2011. pp. 316-317

[55] Coutinho CS, Fermino SAC, Dos Santos BA, Araújo DCL, Da Silva LC. Potential of Hawaii 7996 hybrid as rootstock for tomato cultivars. Bragantia. 2006;**65**(1):89-96. ISSN: 1678-4499

[56] Khah EM, Kalawa E, Mavromatis A, Chachalis D, Goulas C. Effect of grafting on growth and yield of tomato (*Lycopersicon esculentum* mill.) in greenhouse and open-field. Journal of Applied Horticulture. 2006;**8**(1):3-7. ISSN: 0972-1045

[57] Valadez LA. Producción de hortalizas. México: Limusa; 1998. 298 p

[58] Davis AR, Perkins-Veazie P, Sakata Y, López-Galarza S, Maroto JV, Lee SG, et al. Cucurbit grafting. Critical Reviews in Plant Sciences. 2008;**27**:50-74. DOI: 10.1080/ 07352680802053940

[59] Alvarez HJC, Castellanos RJZ, Camacho FF, Aguirre MCL, Rangel LJA, Huitron RMV. Portainjerto y densidad de poblacion sobre la sanidad, rendimiento y calidad de frutos de sandia en lotes con nematodos y *Fusarium*. In: Memorias del XIV. Chapingo Mexico: Congreso Nacional de Ciencias Agronomicas; 2012. pp. 136-137

[60] Alvarez HJC, Castellanos RJZ, Camacho FF, Aguirre MCL, Rangel LJA, Huitron RMV. Comportamiento de la sandía injertada expuesta a suelos con

problemas fitosanitarios y densidades de plantacion. In: Rangel LJA, Raya PJC, Cervantes, OF, Aguirre MCL, Ramirez PJG, Mendoza EM, editors. 1st. Congreso Nacional de Ciencia y Tecnologia Agropecuaria; Roque Guanajuato Mexico; 2012. p. 265-272. ISBN 978-607-96093-1-3

[61] Alvarez HJC, Castellanos RJZ, Camacho FF, Aguirre MCL, Huitron RMV. Injerto de sandia y distanciamiento entre plantas, expuestas a suelo con *Fusarium* y nematodos. In: Memorias del LIX Annual Meeting of the Interamerican Society for Tropical Horticulture; Septiembre 2-6 2013. Santiago de Queretaro, Queretaro Mexico; 2013. p. 172

[62] Alvarez HJC, Castellanos RJZ, Camacho FF, Aguirre MCL, Rangel LJA, Huitron RMV. Uso de porta-injertos en sandia y distanciamiento para el manejo fitosanitario. In: Memorias del 2°. Roque Guanajuato Mexico: Congreso Nacional de Ciencia y Tecnología Agropecuaria; 2014. p. 72. ISSN: 2448-6620

[63] Alvarez HJC, Castellanos RJZ, Camacho FF, Aguirre MCL, Huitron RMV. Porta-injertos y densidades de poblacion para sandia triploide en Michoacan. In: Memorias del 9°. Morelia Michoacan Mexico: Congreso Estatal de Ciencia Tecnología e Innovación; 2014. pp. 595-597

[64] Guan W, Zhao X, Hassell R, Thies J. Defense mechanisms involved in disease resistance of grafted vegetables. HortScience. 2012;**47**(2):164-170 ISSN: 2327-9834

[65] Alvarez HJC, Castellanos RJZ, Aguirre MCL. Ecotipos silvestres de *Carica papaya* y su potencial como portainjertos. In: Memorias del Tercera Reunion Cientifica Interinstitucional sobre Diversidad Biologica. Mayo 18 2018; Ciudad de Mexico. 2018. p. 71

[66] Alvarez HJC, Valdivia RG, Pardo MS. Seleccion de especies de cucurbitaceas con cualidades robustas para la mejora agronomica a traves del injerto. In: Memorias del Tercera Reunion Científica Interinstitucional sobre Diversidad Biológica; Mayo 18 2018; Ciudad de México. 2018. p. 72

[67] Alvarez HJC, Castellanos RJZ, Aguirre MCL. Desarrollo de injertos de papaya: primera etapa. In: Memorias del 11°. Morelia Michoacan Mexico: Congreso Estatal de Ciencia Tecnología e Innovación; 2016. pp. 1-6. ISSN: 04-2016-082917280600-203

[68] Alvarez HJC, Castellanos RJZ, Aguirre MCL. Comparación entre métodos de injerto en *Carica papaya*. In: Memorias del XVII Congreso Nacional y III Internacional de la Sociedad Mexicana de Ciencias Hortícolas; Nuevo Vallarta Nayarit Mexico; 2017 p. 231-234. ISBN: 978-607-37-0763-3

[69] Alvarez HJC, Castellanos RJZ, Aguirre MCL. Estudio preliminar de materiales de papaya injertado. In: Memorias del XVII. Nuevo Vallarta Nayarit Mexico: Congreso Nacional y III Internacional de la Sociedad Mexicana de Ciencias Horticolas; 2017. pp. 214-217. ISBN: 978-607-37-0763-3

Chapter 5

Aspects of the Particular Genetics of Grapes Prolonged for All Horticulture Crops

Vladimir Volynkin, Alla Polulyakh, Svetlana Levchenko, Irina Vasylyk and Vladimir Likhovskoy

Abstract

The modern level of knowledge development in the field of fundamental sciences makes it possible to reliably investigate the processes of evolution. The purpose of our research was to determine the need to establish the existing evolutionary transformations in resistance to abiotic and biotic stress factors of the biosphere in a grape plant, which may be natural for all horticulture crops, and on the other hand, based on the postulate of natural and experimental evolution, to prove the processes of natural evolution as a result of experimental breeding. The results obtained in the study of particular issues of genetics of grapes, based on the existence of general biological regularities, can be prolonged for interpretation, with reference to other horticulture crops. We studied the genetics of grapes, in particular crossability, the inheritance of signs and characteristics, the establishment of regularities in the display of selection value, and heterosis, allowing us to formulate the principles of modeling a new variety. Investigating the process of creating grape varieties that are resistant to biotic factor, it was suggested to consider it from the point of view of the coevolution of the plant and pathogen.

Keywords: grapes, evolution, general and particular genetics, resistance to abiotic and biotic stress factors, crossability, new variety modeling

1. Introduction

Genetic resources of Vitaceae grape family from the various centers of origin and vine-growing regions in the world were taken from one of the oldest Eurasian collections of grapes at the All-Russian National Research Institute of Viticulture and Winemaking (ARNRIVW) “Magarach.” They have been studied on display of genetically determined sign of frost and mildew resistance. It is established that the grape frost and mildew resistance sign in genome was formed and fixed by evolution at separate forms in the various centers of origin. Suggested sources of high frost and mildew resistance of various botanical taxa from the various centers of origin for selection of frost- and mildew-resistant grape varieties are selected also.

Regularities governing the expression of resistance to *Plasmopara viticola (Berk. and M.A. Curtis) Berl. and De Toni*, *Uncinula necator (Schwein) Burrill*, and *Botrytis cinerea Pers.*, the causing agents of mildew, oidium, and gray rot, respectively, and the inheritance of this characteristic in the F_1 progeny were established in grapevine.

The nature of the inheritance of resistance to these pathogens in grapevine seems to be identical despite the fact that they differ in the ecogeographical origin, biological characteristics, and the form of host-pathogen relations (obligate/facultative, pathogenic/saprophyte types). The inheritance of resistance obeys the principle of hypothetic heterosis, which means that the progeny does not contain forms with resistance superior to that of the more resistant parent but does contain forms with resistance surpassing the average level of resistance of the two initial forms. Evidence was also gained in support of the theory suggesting that forms of grapevine with resistance to pathogens emerge only in the course of long-term coevolution of the biological objects. Basic principles of grape breeding were established and confirmed by the findings arising from the search of initial forms of grapevine and by the highlighted regularities governing the inheritance of resistance to pathogens in the crop. Breeding for resistance to pathogens was viewed as a specific target without respect to other desired characteristics since this issue was tackled with an aim to determine breeding and genetic regularities of the process and, for the first time, in terms of the interaction of two biological objects, the host plant and the pathogen, each with its own variation and evolutionary patterns.

The existence of different centers of origin of cultivated plants and grapevine in particular suggests formation of specific biological features in the autochthonous varieties in those centers in the process of evolution. The biological specificity of the autochthonous grapevine varieties was to gain a foothold at the genetic level. At the same time, expression of these traits in the progeny should also be specific. The conducted research has shown that crossing capacity of the Crimean autochthonous grapevine varieties with the varieties and forms of different species and ecogeographical origin has a specific characteristic and proves that crossing capacity can be perceived as an independent genetically determined biological feature.

Global and local climate changes demanded creation of new cultivars, in particular in viticulture. Thus, the problem of breeding new grapevine cultivars that would correspond to the present-day biosphere conditions emerged. As an answer to this, an immunobreeding program "Analogue" was developed at the "Magarach" Institute. The program aims at improving the efficiency of grape breeding and achieving cultivars that would comprise resistance to a variety of pests, diseases, and unfavorable environmental factors and would, at the same time, produce fruit of the quality that would be comparable to the best samples of their international analogues.

The suggested concept of the models is based on 16 most desired selection traits that had been chosen from the existing selection pool, identification of the original forms for crossbreeding, and multivariate analysis. To facilitate cooperation with other research centers, the information obtained was encoded according to the International Organisation of Vine and Wine (OIV)-Bioversity International descriptor.

Using the developed models for the creation of new cultivars will, on the one hand, help solve ecological problems in the vineyards and, on the other, ensure high economic effectiveness of viticulture and winemaking and, at the same time, obtain grape cultivars for different purposes.

2. Materials and methods of investigation

2.1 Materials and methods for gene pool diversity investigation

On the ampelographic collection of the "Magarach" Institute in the field during the years with critical sub-zero temperatures that damage the grape plant, 16 species of the Vitaceae family, 32 varieties of complex interspecific origin, 15 hybrid varieties of *V. vinifera* L. × *V. amurensis* Rupr., 6 varieties of hybrids

V. labrusca L., and 64 samples of the cultivated vine *V. vinifera* sativa D.C., including varieties *V. vinifera* occidentalis Negr., *V. vinifera* pontica Negr., and *V. vinifera* orientalis Negr. and 30 hybrids of *V. vinifera* L. were studied. The study of resistance to frost and agrobiology study of samples were carried out by the method of ampelographic description and agrobiology evaluation of grapes [1]. The degree of damage to annual shoots was assessed on a scale from 0 to 5 on the degree of damage to the phloem, and the analysis of the restoration capacity of the bush on the fruit links and perennial wood was carried out according to the percentage of blossoming shoots on a scale from 0 to 9 points. Estimation of resistance to mildew from 1 to 9 points was carried out according, as well as to frost, to OIV descriptor [2] on the basis of long-term observations (2000–2017) in the years of mildew epiphytotic. A higher score characterized less damage or better restoration ability of the plant, which reflects its higher resistance to stress factors or abiotic (frost) or biotic (mildew) nature.

To study the resistance to mildew, samples of various genetic origins were also selected on the ampelographic collection "Magarach":

- 27 species of the Vitaceae Lindley family, of which 3 are species of the genus *Ampelopsis* Michaux, 2 species of the genus *Parthenocissus* Planch., and 22 species of the genus *Vitis*.
- 198 varieties of complex interspecies origin, which are hybrids of 3 or more species of grapes.
- 18 hybrids of *V. vinifera* L. × *V. amurensis* Rupr.
- 27 hybrids *V. labrusca* L.
- 5 hybrids of *V. riparia* Michx.
- 150 samples of cultivated vine *V. vinifera* sativa D.C. and the European-Asian species (*V. vinifera* L.), which are represented by the varieties of 3 ecogeographical groups: *V. vinifera* sativa convar. Pontica Negr., *V. vinifera* sativa convar. Occidentalis Negr., and *V. vinifera* sativa convar. Orientalis Negr. [3].

2.2 Materials and methods for investigation of host-pathogen relationship

The research was done using a total of 28 hybrid populations of grapevine obtained via hybridization of 31 initial forms, of which 23 and 14 forms entered as the female and the male parents, respectively, and 6 forms were used in both qualities. The genetic resources employed consisted of 1378 hybrid forms belonging to hybrid populations, each containing at least 25 forms [4]. This enabled good reliability of results indicating the inheritance of the characteristic in the progeny [5]. The reliability of the results was confirmed statistically [6]. Initial forms differing in resistance to the causing agents of mildew, oidium, and gray rot (*Plasmopara viticola* Berl. et de Toni, *Uncinula necator* Burr., *Botrytis cinerea* Pers.) were chosen with a view to establish regularities governing the expression of the characteristic in the F_1 progeny. The resistance was assessed by means of a five-point numerical scale, "1" and "5" being the highest and the lowest levels, respectively [7].
The inheritance of resistance in the progeny was analyzed using conventional methodologies [8].

2.3 Materials and methods for crossability investigation

The research was conducted by the Department of Breeding, Genetics and Ampelography of the All-Russian National Research Institute of Viticulture and Winemaking "Magarach" on the field of plant breeding plots located on the southern coast of the Crimea and on the hydroponic culture of the vegetation plot in the period of 2008–2015. The seedlings were grown on gravel in hydroponic channels. For the cultivation of grape seedlings, we used chemically inert fraction of diorite aggregate sized 3–5 mm, which is a good conductor of the nutrient solution to the plant roots that keeps water on its surface well, serves as a support for the roots, and provides good aeration of the root system. For the growth of grape seedlings, we used hydroponic solution of variable composition [9, 10].

During the investigation period, 43 combinations of cyclic crosses were performed. As female initial forms, we used varieties and hybrids with female flower type: 4 Crimean autochthonous grape varieties ('Kefessia', 'Krona', 'Sary Pandas', 'Kok Pandas') and 2 interspecific hybrids ('Muscat Jim' and Magarach № 31-77-10). As male initial forms, we used 3 Crimean autochthonous grape varieties ('Shabash', 'Kokur Belyi', 'Gevat Kara'); 3 autochthonous varieties from other centers of origin of grapes ('Rkatsiteli', 'Chardonnay', 'Sauvignon vert'); 2 varieties of the Institute of "Magarach" selection ('Bastardo Magaraci' and 'Rubinovyi Magaracha'), derived by crossing within the species of *Vitis vinifera*; and 11 interspecific varieties ('Podarok Magaracha', 'Citronnyi Magaracha', 'Aurora Magaracha', 'Pervenets Magaracha', 'Riesling Magaracha', 'Spartanets Magaracha', 'Krasen', 'Antey Magarachsky', 'Alminskyi', 'Granatovyi Magaracha', and 'Pamyati Golodrigi').

We studied hybrid seed formation, fully formed seed (seeds with a viable germ and well-developed endosperm) formation, seed germination, and development of seedlings, including vigorous ones as they are most viable and potentially more productive. The reliability of the results was confirmed by experimental data processing using mathematical and statistical processing methods [4, 5, 6, 8].

2.4 Materials and methods for plant breeding modeling investigation

Based on the formulated research objectives, the expressiveness of traits in the researched grapevine gene pool was studied with the purpose of their subsequent use as gene resource pool and in hybrid progeny. Acquisitions were described according to the OIV Official List [1].

Fourteen grapevine cultivars as initial forms along with 3704 hybrid seedlings obtained from 40 combinations of cyclic crossings were studied. The study and selection of seedlings were conducted based on the methodology developed by the Department of Breeding, Genetics and Ampelography of the "Magarach" Institute [11]. The degree of the manifestation of traits was coded [12]. The obtained primary data on the degree of manifestation of traits in original forms and hybrid seedlings was processed by methods of mathematical statistics [6]. Combinational ability, breeding value, and hypothetic heterosis [4, 8, 13] were estimated.

3. Genetic diversity in grape family Vitaceae

Genetic resources of cultivated plants with valuable biological characteristics, used for food production, ensure stable development and functioning of ecologically safe agricultural sector in conditions of constant changes in natural and climatic factors and social circumstances. Population growth and economic development of countries make significant changes in the living conditions of all organisms and ecological systems of the planet [14].

Mobilization of genetic resources of grapes in ampelographic collections, as shown by scientists of many countries [15–17], plays an important role in the conservation and use of the gene pool of grapes. One of the important tasks of collecting and preserving the gene pool of the genus *Vitis* in many countries of the world is the preservation of local varieties of grapes that are exclusively part of the natural heritage and do not grow in other vine-growing regions [18, 19]. Most of the local and uncommon varieties of grapes are now preserved only through collections [20]. It is known that the local assortment of grapes has been formed for a long time under certain conditions of a specific region and has a number of valuable features [21]. The study of the genetic resources of grapes collected on ampelographic collections from different centers of origin of grapes will allow us to identify new sources for breeding, to form an idea of the evolution of the grape culture.

Genetic resources of the grape of the "Magarach" Institute are collected in an ampelographic collection, one of the largest and oldest collections of grapes in the world, which was noted at international scientific forums and Internet sites [22]. The beginning of the formation of the collection dates back to 1814 and the existence of the "Magarach" Institute (1928) and coincides with the period of the beginning of collecting varieties of vine in Europe. The uniqueness of the collection of the Magarach Institute is that over two centuries it was replenished with samples from various vine-growing regions of the world, Europe, Asia, Africa, and North America, and most of them have survived to this day.

Proceeding from the tasks of general and private ampelography as a branch of botany, the extensive gene pool of grapes in the collection of the "Magarach" Institute has been studied and studied for the purpose of classifying samples, based on their origin in different centers and their evolutionary variability. Continuous work in this direction made it possible, in particular, to clarify the classification of the Eurasian species of grapes *Vitis vinifera* L., isolating subspecies, varieties within the subspecies of wild forest grapes, ecogeographical groups, and groups and types of cultivars.

Samples of the ampelographic collection were studied not only as separate botanical species, differing in morphological and morphometric characteristics, for the purpose of forming botanical taxons but also as genetically formed and fixed in the genotype biological specificity, in particular resistance to stress factors of the biosphere.

The analysis of the safety indices of the main and replacement buds, the degree of frost damage of the annual vine, as well as the restoration ability of the specimens of the samples of the collection showed that in the species of the genus *Parthenocissus* Planch., 100% safety of the main and replacement buds was recorded. Also high frost resistance was noted in species of the genus *Vitis* such as *Vitis* Linn., *Vitis amurensis* Rupr., North American species *Vitis cinerea* Arnoldi, *Vitis longii* Br., *Vitis riparia* Michx., and *Vitis solonis* Planch. Preservation of the main buds in these species was 95–98%, replacing 98–100%.

Analysis of varieties of grapes of interspecies origin showed that in hybrid varieties *Vitis vinifera* L. × *Vitis amurensis* Rupr., safety of the main buds was 1–25%, replacing 30–65%. The lesions of the annual vine were not detected. Varieties showed a good restorative capacity (**Figure 1**): on the fruit links, the percentage of developed shoots was 25–75% and on the bush stem from 50 to 100%. However, half a century of experience with the hybrids *Vitis vinifera* L. × *Vitis amurensis* Rupr. did not reveal genotypes combining a high level of frost resistance (−40°C) with signs of the quality of berries of cultured grapes. Created genotypes exhibit an average level of frost resistance in comparison with parental species (−25–26°C

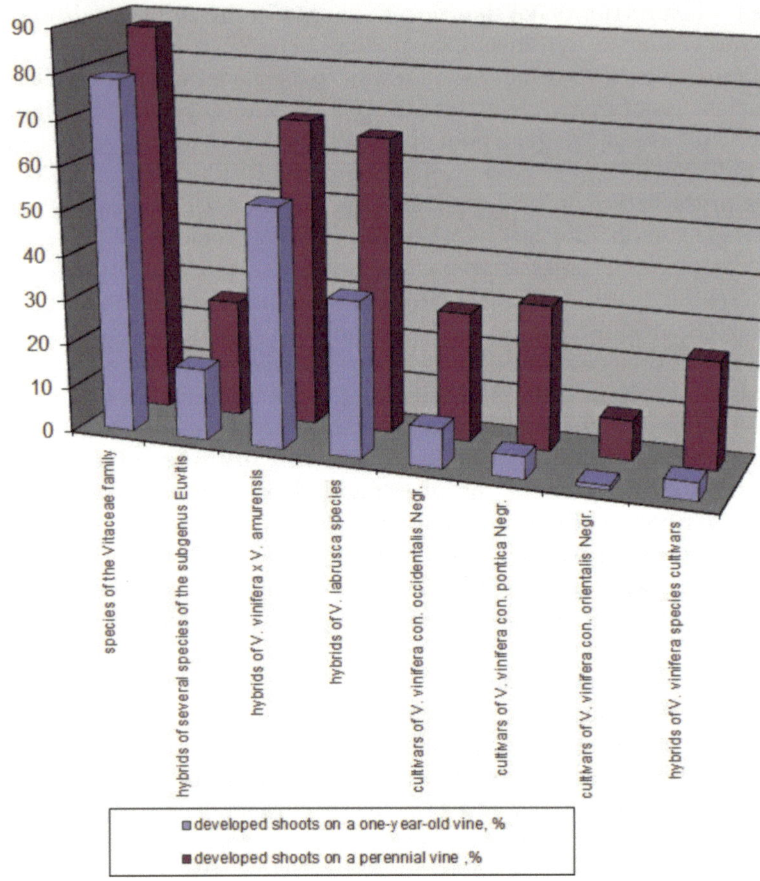

Figure 1.
Frost resistance of the grape samples of the ampelographic collection "Magarach" various origins.

table varieties and 27–28°C technical varieties). Similar results are given by crosses with frost-resistant American species.

Increased frost resistance is also possessed by some cultural representatives of the species *Vitis vinifera* L. due to the high ecological plasticity of the species. Under the influence of natural and artificial selection in various regions of formation, some varieties of cultured grapes have acquired adaptive properties with respect to low negative winter temperatures. Such varieties are of particular value for breeding, because unlike wild representatives of the genus *Vitis*, they do not carry the heredity of wild genotype, which reduces the quality of grapes and wine in hybrid forms in the heredity. In general, analyzing the safety indices of the main and replacement buds, the degree of frost damage of the annual vine, as well as the restoration ability of *Vitis vinifera* L. varieties, according to the degree of frost resistance, they can be divided into three groups:

- Varieties with increased resistance to frost. These are the cultivars *V. vinifera occidentalis* Negr. ('Aligoté', 'Cabernet-Sauvignon', 'Merlot', 'Riesling', 'Italia', 'Riesling Rhine', 'Sauvignon Green'), cultivars *Vitis vinifera pontica* Negr. ('Rkatsiteli', 'Saperavi'), and cultivars-hybrids *Vitis vinifera* L., parental forms of which have increased resistance to frost: 'Odesskii chernyi' ('Alicante Bouschet' × 'Cabernet-Sauvignon') and 'Sukholimanskii belyi' ('Chardonnay' × 'Plavay').

- Varieties with poor frost resistance. These are the varieties *Vitis vinifera pontica* Negr. ('Fetyaska', 'Furmint', 'Chausch white') and cultivars-hybrids *Vitis vinifera* L. ('Odesskii souvenir', 'Sverkhrannii bessemyannyi', and 'Suruchenskii belyi').

- Not frost-resistant varieties. These are cultivars of the ecogeographical group *Vitis vinifera orientalis* Negr. ('Asma', 'Karaburnu', 'Muscat Alexandria', 'Muscat white') and cultivars-hybrids *Vitis vinifera* L., created on the basis of cultivars of the eastern ecogeographical group: 'Irsai Oliver' ('Pozhoni white' × 'Pearl Saba'), 'Beauty Tsegleda' ('Chasla white crispy' × 'Chasla pink royal'), 'Muscat Hamburg' ('Muscat Alexandrian' × 'Frankenthal'), and others.

A comparative analysis of the regenerative capacity of bushes after the defeat of extreme winter frosts by groups of different origin showed (**Figure 1**) that the wildest species of the *Vitaceae* L. family are the highest adaptive capacity, the less the hybrids *Vitis vinifera* L. × *Vitis amurensis* Rupr. 53.8% of bloomed shoots on fruit segments and 68.8% of blossom shoots on perennial wood. Below these, the *Vitis labrusca* L hybrids: 35.0 and 66.2%, respectively. Varieties of complex interspecific origin have an average of 12.5 and 16.0% of blossoming shoots on fruit segments, but in terms of the number of restored shoots, they exceed the local varieties *Vitis vinifera occidentalis* Negr. and *Vitis vinifera pontica* Negr.

Mildew resistance studies have shown that 27 species of the Vitaceae Lindley grape family possess the highest possible. The assessment of the degree of resistance was 9 points (**Figure 2**). Also the hybrids of *Vitis riparia* Michx. possess the maximum resistance to mildew among which are 'North White', 'Taiga Emerald', and others with an average resistance score of 9.

In cultivars-hybrids *Vitis vinifera* L. × *Vitis amurensis* Rupr., average resistance is 8.6 points. The degree of resistance is 9 points for 'Aksay', 'Golubok', 'Far Eastern Tikhonov', 'Dyushes', and 'Zarya Severa'; the degree of resistance is 7 points for cv. 'Anushayut', 'Bruskam', 'Karin', 'Kunlean', and 'Russian early'; and the degree of resistance is 5 points for cv. 'Cherry Early', 'Dniester Pink', and 'Negru de Yaloven'. In the cultivars-hybrids of *Vitis labrusca* L., the average score of resistance is 8.6, among them are resistant cultivars (9 points): 'Alpha', 'Lyathes', 'Romulus', 'Lusay'l', and 'Isabella'. The cultivars 'Madella', 'Menu', and 'Lydia' had 7 points.

Less resistant varieties of complex interspecific origin have an average resistance score of 8.06. Varieties with maximum resistance of 9 points include 'Antey Magarachsky', 'Bianka', 'Aurora', 'Golubok', 'Golden Muscat', 'Podarok Magaracha', and 'Tair'. Relatively resistant varieties with a degree of resistance of 7

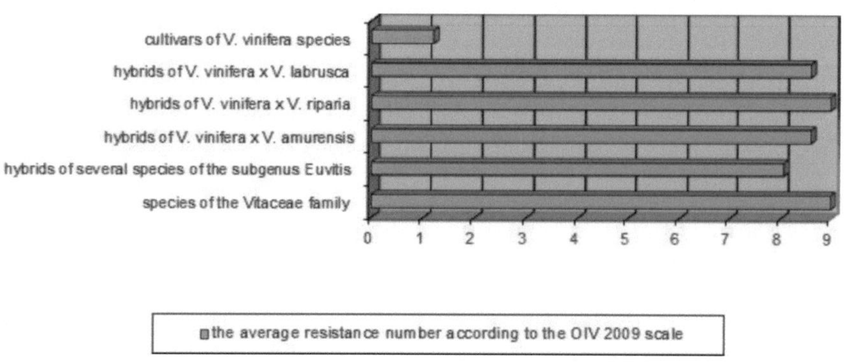

Figure 2.
Mildew resistance of the grape samples of the ampelographic collection "Magarach" various origins.

points include 'Agavam', 'Armalaga', 'Artages', 'Biruinz', 'Villar Blanc 12-375'), 'Vostorg', and 'Citronny Magaracha'. For the varieties 'AsmaMagaracha', 'Victoria', 'Druzhba', and 'Skorenskiy red', the degree of resistance was 5 points.

The average resistance score for cultivars of *V. vinifera* sativa D.C. is 1.2. Most cultivars of *V. vinifera* sativa D.C. are not resistant to mildew as the evolutionary pathogen and variety samples of this species were developed in different centers of origin of the grapes and were geographically isolated. These are such varieties as 'Cardinal', 'Kishmish black', 'Goruli mtsvane', 'Kok habah', 'Murza raisin', and 'Artin zerva'.

Relative resistance to the pathogen is noted in some local varieties like 'Chausch black', 'Chilar', 'Chinuri', 'Chol Ber', 'Shaani white', 'Shaani black', 'Aygesard', 'Bayan shirei', 'Varyushkin', 'Tresso black', 'Vereya', 'Chersonesus', 'Hindogny', and 'Tsolikouri', whose resistance levels were 7 points, and varieties 'Chausch white', 'Albilo Crimean', 'Mattrassa', 'Misket Cherven', 'Sabash', whose resistance level to mildew was 5 points.

Botanical diversity of grapes culture, reflected in the classification and taxonomy of varieties and forms of the grape Vitaceae family, reflects the process of natural evolution and natural and artificial selection. Scientists in various countries in the eighteenth to twentieth centuries carried out a grandiose work that made it possible to differentiate representatives of the Vitaceae family in botanical taxons. The conducted researches established that all this botanical variety of the Vitaceae family was formed in different centers of origin of the grape culture on the planet. Depending on the abiotic conditions in the centers of origin of culture, a genome was formed in the form of individual botanical taxa in particular within genera and species differentiated in resistance to abiotic stress factors. Biotic conditions in these same centers of origin formed the genome in the form of the same individual botanical taxa, within the genera and types of grapes as a host plant, differentiated by resistance to biotic stress factors and the pathogen gene in the process of conjugate evolution. In the end result, samples of grapes were formed in each separate center of origin of the culture, differentiated not only according to botanical characteristics but also on a set of biological characteristics, in particular, resistance to biotic and abiotic stress factors of the biosphere, as frost and mildew resistance.

Speciation, specifically the formation of the species *Vitis vinifera* L., is directly related to the existence of wild forest grapes, belonging to the relics of Eurasia. The studies established a significant difference, including morphological and morphometric features and molecular genetic markers, between forms of wild forest grapes from various regions of Eurasia. Consequently, it may be considered necessary to continue these studies in order to isolate in the centers of origin of the culture of the grapes individual foci or subcenters of origin.

4. Plants as host and pathogen coevolution

Grapevine, like any plant species defined botanically as an independent hierarchical unit, has its own habitat [23], which, in turn, embraces individual foci of origin [24].

For each agricultural crop, including grapevine, both habitats of varieties in commercial cultivation and foci of origin of different botanical forms are determined [3]. Such varieties make up the commercial assortment of a crop, and botanical forms constitute the family of Vitaceae [25, 26]. The objective of this research was to establish genetically determined regularities governing resistance to pathogens in grapevine based on accessions from different foci of origin and newly bred forms obtained via sexual hybridization [27].

A considerable amount of research into the capacity of hybridization in grapevine and the inheritance of characteristics, including resistance to mildew, oidium, and gray rot in the F_1 progeny, has been done in the "Magarach" Institute. Regularities governing the expression of the biological specificity of the capacity of hybridization or the inheritance may be either general or particular; one can speak about either general biological regularity, the nature of the combining ability of two definite forms of grapevine, or the possibility to observe a certain degree of the expression of the characteristic in the F_1 progeny.

Ecogeographical foci of origin of initial forms, the interrelationships of these foci with the origin of individual groups of forms which may be grouped into botanical taxa, the degree of genetic relatedness of forms obtained via inbred and distant hybridization, the resistance of initial forms to a given pathogen, the presence of forms with different levels of resistance (assessed by use of a numerical scale) in the progeny, and the average resistance scores across the population may serve as the key elements in considering different approaches in attempting to examine the outcome of breeding varieties and forms of any crop, including grapevine, for resistance to pathogens. All this helps investigate the genetics of the inheritance of resistance. The results obtained may be interpreted from a standpoint of the abovementioned approaches, e.g., a genetic formula of initial forms may be described, these initial forms may be assessed as sources and/or donors of the characteristic in question, and the effectiveness of the transmission of the characteristic during hybridization with definite individual forms may be determined. The influence of abiotic factors on the effectiveness of the capacity of hybridization in grapevine and the effectiveness of the transmission of characteristics to the progeny has not been studied so far.

In order to develop breeding programs, including those aimed to achieve grape varieties resistant to pathogens, we need to know the biological peculiarities of the plant and the specificity of desired characteristics. Breeding grape varieties for resistance to biotic factors may require consideration of the biology and ecogeographical origin of the pathogens in question.

Though the causing agents of mildew, oidium, and gray rot are all fungi, they do differ biologically. *Plasmopara viticola* and *Uncinula necator*, the causing agents of mildew and oidium, respectively, are capable to grow only on grapevine as the obligate pathogens, while *Botrytis cinerea*, the causing agent of gray rot, is a cosmopolite, polyphage, and saprophyte. Biological differences of the pathogens suggest that the inheritance of resistance to them as a genetically determined characteristic is governed in grapevine by specific regularities.

Nevertheless, the results obtained indicate that the nature of the inheritance of resistance to these pathogens in grapevine seems to be identical despite the fact that they differ in biological characteristics and the form of host-pathogen relations (obligate/facultative, pathogenic/saprophyte types). Neither was the inheritance of resistance affected by the feeding patterns of the pathogens. Therefore, in developing programs of breeding grape varieties with resistance to the pathogens, their biological peculiarities need not be taken into account, and also attempts may be made to combine resistance to all three diseases in one genotype. Results highlighting these conclusions are shown in **Tables 1** and 2.

The data in **Table 1** indicates that resistance of initial forms and their combining ability are the principal factors underlying resistance of the progeny. Another finding is that the outcome is not affected by using initial forms with better resistance as the male or the female parent.

The possibility to achieve resistance to a set of pathogens has been viewed differently in the beginning of grape breeding activities. Some researchers favored the possibility of breeding simultaneously for a set of desired characteristics, while

Resistance scores of				
Initial forms		F_1		
Female parent	Male parent	Mildew	Oidium	Gray rot
5	3	4.0	4.0	3.9
-”-	2	3.4	3.7	2.8
-”-	1	2.6	2.7	2.8
4	4	3.7	4.0	3.8
-”-	3	3.5	3.7	3.5
-”-	2	3.4	3.4	3.1
-”-	1	3.0	3.1	2.3
3	5	4.3	3.9	4.1
-”-	4	3.8	3.6	3.7
-”-	3	3.4	3.1	3.0
-”-	2	3.3	2.9	2.7
-”-	1	2.5	2.2	2.3
2	5	4.4	3.3	3.6
-”-	4	3.6	3.4	3.1
-”-	3	3.4	2.9	3.0
-”-	2	2.1	2.3	2.3
-”-	1	2.0	1.9	1.8
1	4	3.6	3.4	3.3
-”-	3	2.9	3.0	2.3

Table 1.
Effect of resistance of initial forms of grapevine to pathogens on the inheritance of the characteristic in the F_1 progeny.

others suggested breeding for resistance to a certain pathogen followed by subjecting the intermediate cross to saturation hybridization, which added resistance to the second element of the set, etc., in a stepwise fashion. The advent of complex infection backgrounds has enabled the breeder to assess the newly bred genetic resources for resistance to a set of pathogens and to confirm that hybrids with resistance to the set of the pathogens in question have been achieved provided the appropriate choice of initial forms. Data in **Table 2** shows regularities governing the inheritance of resistance to the set of the pathogens in question in grapevine.

Table 2 highlights the inheritance of resistance to three pathogens in grapevine, in which pathogens differ in their biological characteristics and host-pathogen relations and are capable, during the vegetation period, to compete at the infection stage. Nevertheless, the fact that a number of populations exist, such as 'Nimrang' × 'Magarach 124-66-26', 'Plechistik' × 'Magarach 124-66-26', and 'Plechistik' × '*Antaeus magarachskii*', with "3" as the average score of resistance to the three pathogens across the population supports the suggestion that simultaneous breeding for resistance to a set of pathogens is possible.

In a study of regularities governing the inheritance of resistance to various pathogens in grapevine, both theoretical peculiarities and practical aspects of their use need to be taken into account. It appears that they should be viewed in terms of the possibility to achieve definite forms of grapevine with resistance to a set of

Cross-combination	Resistance scores of F_1		
	Mildew	Oidium	Gray rot
'Nimrang' × 'Magarach 124-66-26'	3.1	2.2	3.1
'Madeleine angévine' × 'Magarach 124-66-26'	4.8	3.0	3.1
'Queen of vineyards' × 'Magarach 124-66-26'	2.8	2.9	3.5
'Plechistik' × 'Magarach 124-66-26'	2.9	3.2	2.6
'Tachly' × 'Seyve Villard 20366'	3.3	3.0	2.2
'Seyve Villard 20365' × 'Italia'	4.9	3.3	3.7
'Magarach 4-68-25' × 'Krymskaya Zhemchuzhina'	4.0	2.8	4.0
'Seyve Villard 12283' × 'Sverkhrannii bessemyannyi Magaracha'	4.3	3.9	3.0
'Kefessia' × 'Antaeus magarachskii'	2.5	3.3	2.1
'Plechistik' × 'Antaeus magarachskii'	2.6	3.4	2.8

Table 2.
Comparative inheritance of resistance to different pathogens of grapes.

pathogens and to reproduce the results achieved in repeated crossings provided the use of the same initial forms. The understanding of such regularities remains incomplete since grapevine is a complex heterozygous organism. However, for a number of individual characteristics, including resistance to definite pathogens, certain results have been obtained and expressed as the scores of a universal numerical scale.

Natural selection reigns in the plant kingdom, and new forms of a given crop may emerge and become established in plant communities. It is highly probable that mutations or events of natural hybridization give rise to such forms. New forms of both types arise due to a change in forms of a crop within one botanical species, in which the forms, as a rule, grow geographically isolated from other species of a given crop. That is why the richest diversity of grapevine known prior to the 1860s when the first purposeful crossing of two forms of the crop took place should be viewed as the outcome of formication within individual species which, in turn, come from the European-Asian, North American, South American, or East Asian foci of origin of grapevine.

Over centuries of coevolution, grapevine, phylloxera, and the causing agents of mildew, oidium, and gray rot have been subject to slow permanent changes, leading to an increased virulence of the pest and the pathogens and to the development of defense reactions by the crop. This long-term process has brought about the emergence of stable forms of grapevine. Scientists stated that resistant forms emerge as a result of coevolution of the host plant and the pathogen, and that is why they should be searched for in the birthplace of the latter [28–30].

The successful implementation of breeding programs aimed to achieve varieties of the future envisages, as an indispensable prerequisite, the involvement of the global genetic resources of an agricultural crop into the breeding process. The diversity of the global genetic resources suggests the possibility in combining properties inherent in species formed at different ecogeographical conditions in one genotype.

The use of a species for breeding purposes means that the entire diversity of a species in a habitat needs to be collected and studied. Forms should be collected in sufficiently large quantities to enable studies of the variation of the species for characteristics in question and the ability of forms to transmit them to the progeny.

Breeding new generations of grape varieties distinguished for excellent quality of the fruit and good yielding capacity and also resistant to major diseases and pests of the crop is an important task throughout the viticulture world. In the absence of artificial infection backgrounds which help the breeder test resistance of the grape genetic resources to definite pathogens, resistant varieties and forms can be revealed against natural epiphytotics.

The grape plant, as a biological object, exists and develops in the biosphere, being in permanent contact with and affected by biotic and abiotic factors. Careful consideration is given to the effects of biotic factors on development of a plant organism, including grapevine, due to the specific nature of the phenomenon, the essence of which lies in the fact that two biological objects, the host plant and the pathogen, change as affected by each other and abiotic factors.

The fact that variation of biological objects is governed by the abovementioned regularities enables evolutionary models to be built. Besides, and this is of no less importance, this knowledge can be applied in purposeful creation of forms of a crop which possess a desired set of useful characteristics, including resistance to pathogens. The expression of such characteristics must be high enough to allow cultivation of these forms in the field without appropriate protection measures.

Grapevine may be attacked by pests and diseases whose causing agents are of different natures: viruses, bacteria, and fungi. The biology and the life cycle of the fungi *Plasmopara viticola*, *Uncinula necator*, and *Botrytis cinerea*, the causing agents of mildew, oidium, and gray rot, respectively, have most exhaustively been studied. The development of grapevine as affected by the former, the reaction of the grape plant to damage, and the degrees of resistance of forms, varieties, and different species of the crop to these pathogens have also been studied in detail.

Different degrees of resistance to the causing agents of the abovementioned diseases in individual grape varieties suggest that purposeful breeding of varieties is possible which combines resistance to the set of the pathogens in question and other desired economical characteristics in their genotypes. To this end, we need to know regularities governing the capacity for hybridization of various initial forms and the inheritance of characteristics, including resistance to pathogens, in the F_1 progeny. Since grapevine is a vegetative propagated crop, we may limit our studies of the inheritance of characteristics to the F_1 progeny, and the transmission of characteristics to the progeny is governed both by the capacity for hybridization of initial forms and their combining ability, which determine the transmission of a characteristic to the progeny and the degree of its expression in the F_1 hybrids.

A number of regularities governing the expression of resistance to *Plasmopara viticola*, *Uncinula necator*, and *Botrytis cinerea*, the causing agents of mildew, oidium, and gray rot, respectively, and the inheritance of this characteristic in the F_1 progeny were established in grapevine as a result of our long-term research aimed to breed for resistance of the crop to these fungal pathogens. The nature of the inheritance of resistance to these pathogens in grapevine seems to be identical despite the fact that they differ in biological characteristics and the form of host-pathogen relations. The inheritance of resistance obeys the principle of hypothetic heterosis, which means that the progeny does not contain forms with resistance superior to that of the more resistant parent but does contain forms with resistance surpassing the average level of resistance of the two initial forms.

Evidence was also gained in support of the theory suggesting that forms of grapevine with resistance to pathogens emerge only in the course of long-term coevolution of the biological objects. That is why the search for forms of grapevine with resistance to the given pathogens, which remains important today, should be based on that principle.

In conclusion, it should be mentioned that basic principles of grape breeding were established and confirmed by the findings arising from the search of initial forms of grapevine and by the highlighted regularities governing the inheritance of resistance to pathogens in grapevine. Breeding for resistance to pathogens was viewed as a specific target without respect to other desired characteristics since this issue was tackled with an aim to determine breeding and genetic regularities of the process and, for the first time, in terms of the interaction of two biological objects, the host plant and the pathogen, each with its own variation and evolutionary patterns.

5. Crossability of plants as a genetic determined sign

In this stage of scientific knowledge and expertise, the concept of evolutionary development of the flora on Earth is based on the existence of certain centers of origin of cultivated plant species, including grapes [24]. A fundamental contribution to the development of the theory of centers of origin was made by Vavilov [31]. Negrul has further developed this theory [32], highlighting the ecological and geographical areas forming the autochthonous grape varieties. These varieties form separate ecogeographical groups of autochthonous grape varieties. The "Black Sea" area is one of such ecogeographical zones, where autochthonous grape varieties are formed [33]. Within the bounds of this zone, modern scientists have identified specific subcenters of origin of the grapes, one of them being Crimea [21, 34], where relict endemic wild forest vines exist to date. This resulted in the formation of a specific set of autochthonous grape varieties with characteristic biological features in the Crimea, due to which they are used in the breeding programs developed by the "Magarach" Institute [3].

It should be noted that the process of breeding new varieties, as a rule, involves analysis of the combining ability of the initial forms with regard to the inheritance of the degree of trait expression in the progeny. At the same time, it has to be kept in mind that in the course of generative hybridization, the formation of a germ and seeds, along with hybridization efficiency, differs between various initial forms, and we can assume that this is a genetically predetermined feature. Therefore, the autochthonous grape varieties of the Crimea are characterized by specific crossability that has become part of their DNA in the process of natural evolution. The present work focuses on the analysis of this phenomenon.

It is known that the majority of the Crimean local grape varieties has a functionally female flower type and is, therefore, not resistant to the biotic and abiotic environmental factors. This affects the stability of fertilization and vine yielding capacity. It is possible to increase the adaptive capacity of the Crimean local grape varieties not only by the selection method but by the hybridization method as well. It is worthwhile to evaluate the hybridization efficiency of the Crimean autochthonous grape varieties, which, first and foremost, require determining the parameters, based on which such studies may be performed. The formation of a new genotype of a plant consists of two principal stages: (1) hybridization, to include formation of berries and seeds, and (2) the development of seedlings and of their vegetative propagation later on.

According to the earlier data, the berry formation in different breeding combinations is characterized by biological traits of the initial female plant, while it is not dependent on the male form, which has been supported by the number of berries per bunch of a cross-combination. Therefore, there is no need to consider the number of bunches and berries per bunch in the further study of grape crossing capacity. When assessing crossing capacity, it is more important to pay attention to the number of the formed seeds, including the fully formed ones, as compared to the number of berries with formed seeds, and to the number of seedlings obtained in the end of the process, including the vigorous ones.

8096 hybrid seeds were obtained as a result of the crosses. As can be seen from the data presented in **Table 3**, which characterize the extreme contrasts among the obtained results, the greatest number of seeds was formed in combinations with 'Muscat Jim' as female form, from 853 to 546 seeds, and the minimum number was formed in the combination of 'Krona' × 'Krasen' (3 seeds).

On average, fully formed seeds accounted for 85%, while the number of fully formed seeds in 'Krona' variety in three cross-combinations reached up to 100%. Less than mean value, from 80 to 46% of fully formed seeds were formed in almost all backcrosses involving 'Kok Pandas' as the female form. Above-average percentage of fully formed seeds was observed in populations of 'Kefessia' × 'Rubinovyi Magaracha' and 'Kefessia' × 'Bastardo Magaraci', 97% and 98%, respectively. Seed germination analyses revealed the lowest percentage in cross-combinations of 'Kok Pandas' with 'Sauvignon vert' (32%), 'Chardonnay' (41%), and 'Rkatsiteli' (46%). Seed germination in combinations of 'Kok Pandas' with interspecific varieties of 'Aurora Magaracha' and 'Riesling Magaracha' made 60%. The maximum viability was observed in the seeds of 'Sary Pandas' variety in combinations with interspecific male forms of 'Citronnyi Magaracha' (82%) and 'Riesling Magaracha' (78%). A lower percentage of germination was observed in cross-combinations of 'Sary Pandas' × 'Sauvignon vert' (76%) and 'Sary Pandas' × 'Chardonnay' (73%). The average seed germination value for all the populations was rather high and reached up to 51.2%.

As a result, we were able to grow 4143 seedlings, out of which only 258 were vigorous (16.7%). Of particular interest in terms of this indicator was the combination of 'Krona' × 'Alminskyi' (17%) and 'Krona' × 'Rubinovyi Magaracha' (15%). Of similar powerful growth force were seedlings in the populations of Magarach № 31-77-10 × 'Gevat Kara' (14%) and Magarach № 31-77-10 × 'Kokur White' (13%). We were not able to single out any vigorous seedlings in combinations of 'Sary Pandas' × 'Chardonnay', 'Sary Pandas' × 'Pervenets Magaracha', 'Kok Pandas' × 'Aurora Magaracha', 'Kok Pandas' × 'Spartanets Magaracha', 'Kok Pandas' × 'Citronnyi Magaracha', and 'Kok Pandas' × 'Rkatsiteli'. The average number of vigorous seedlings in all populations was 6.2%.

The crossability efficiency of the grapes as to the formation of fully formed seeds and their germination capacity with the resulting vigorous seedlings is determined primarily by the use of a particular female form in a cross and its ripening period. In the present study all the female forms had an average ripening period, which provided a sufficiently high percentage of germination. To analyze the effect on the viability of the offspring, we took six initial female forms, which were cyclic crossed with male forms, and grouped them into unified complexes (**Table 4**). The number of seeds produced in these crosses fluctuated significantly from 484 seeds for 'Krona' to 2668 seeds for 'Kok Pandas'. The average sample value was 188.3 of hybrid seeds, while the coefficient of variation for the populations made 38 of seeds. At the same time, the percentage of fully formed seeds out of their total number in cross complexes fluctuated insignificantly, from 87.3% (for 'Sary Pandas') to 95.2% (for 'Kefessia'), with the exception of 'Kok Pandas', which showed 67.9%, as evidenced by the excess of the variation coefficient. The mean value for seed germination capacity in all the complexes made 58.9%. The greatest number of vigorous seedlings was observed in female forms of Magarach № 31-77-10 (13.5%), 'Krona' (10.2%), and 'Muscat Jim' (10%). 'Kok Pandas' produced the lowest percentage of vigorous seedlings that amounted to 1.1%. In all the populations, the vigor of seedling growth varied considerably, with its rate reaching up to 78.7%.

From the analysis of the obtained data, we may conclude that 'Krona' produced the least number of hybrid seeds that came up to 484, 10% of them produced vigorous seedlings, while 'Kok Pandas', although having formed the highest number

Crossbreeding combinations		Number of hybrid seeds	Fully formed seeds, %	Germinating capacity of hybrid seeds, %	Number of seedlings in a combination	Number of vigorous seedlings	Vigorous seedlings, %
Female form ♀	**Male form ♂**						
'Muscat Jim'	'Kokur Belyi'	853	95	47	403	47	12
'Muscat Jim'	'Shabash'	549	95	46	250	18	7
'Kok Pandas'	'Sauvignon vert'	478	48	32	153	1	1
'Kok Pandas'	'Aurora Magaracha'	376	93	60	224	1	1
'Magarach № 31-77-10'	'Kokur Belyi'	216	81	48	104	14	13
'Kefessia'	'Rubinovyi Magaracha'	161	97	63	101	5	5
'Kefessia'	'Bastardo Magaraci'	65	98	65	42	3	7
'Sary Pandas'	'Chardonnay'	88	98	73	64	0	0
'Sary Pandas'	'Citronnyi Magaracha'	33	55	82	27	1	4
'Krona'	'Rubinovyi Magaracha'	71	94	56	40	6	15
'Krona'	'Citronnyi Magaracha'	48	88	58	28	4	14
'Krona'	'Alminskyi'	10	90	60	6	1	17

Table 3.
Crossability efficiency of the Crimean autochthonous varieties with varieties of different origins.

Female form, ♀	Hybrid seeds, pieces	Fully formed seeds, %	Germinating capacity of hybrid seeds, %	Vigorous seedlings, %
'Magarach № 31-77-10'	768	90.0	50.0	13.5
'Krona'	464	93.9	61.5	10.2
'Muscat Jim'	2207	94.7	46.0	10.0
'Kefessia'	1208	95.2	60.1	5.2
'Sary Pandas'	781	87.3	68.9	4.8
'Kok Pandas'	2668	67.9	50.0	1.1
$\bar{x}$	188.3	87.1	58.9	6.1
σ	214.6	14.9	10.2	4.8
V	38.0	17.1	17.3	78.7

Table 4.
The effect of maternal forms on crossability.

of fully formed seeds amounting to 2668, showed the smallest percentage of vigorous seedlings (1.1).

In combination of crosses of *V. vinifera* with complex interspecific varieties chosen for the analysis, when different female varieties were used in hybridization with either the same male variety or with different ones, we were able to obtain a sufficiently high number of seeds. For the crossability of the male form analysis, the initial varieties were grouped into complexes, taking into account all the indicators resulting from the hybridization of one male form with all the female forms.

As can be seen from the data, the mean value for the obtained hybrid seeds made 116 (**Table 5**). The lowest number of seeds was obtained in cross-combinations, where 'Podarok Magaracha' was the male form (18 seeds). The greatest number of

Male form, ♂	Hybrid seeds, pieces	Fully formed seeds, %	Germinating capacity of hybrid seeds, %	Vigorous seedlings, %
'Alminskyi'	72	94.0	64.0	11.0
'Podarok Magaracha'	18	72.0	56.0	10.0
'Pamyati Golodrigi'	52	93.0	61.4	9.5
'Granatovyi Magaracha'	62	89.5	59.5	6.0
'Antey Magarachsky'	155	90.5	58.0	5.0
'Aurora Magaracha'	422	93.0	62.5	5.0
'Krasen'	168	90.3	62.3	4.0
'Riesling Magaracha'	754	95.3	61.7	3.7
'Citronnyi Magaracha'	456	77.3	63.3	3.5
'Spartanets Magaracha'	534	82.0	58.5	3.5
'Pervenets Magaracha'	209	69.5	60.5	1.0
$\bar{x}$	116	86.4	61.2	5.5
σ	136	13.2	8.0	4.6
V	117	15.3	13.1	83.6

Table 5.
The effect of paternal forms of interspecific origin on the crossing capacity.

seeds was observed in crosses with 'Riesling Magaracha' (754 seeds). The number of fully formed seeds out of their total amount was high enough and ranged from 69.5% for 'Pervenets Magaracha' to 95.3% for 'Riesling Magaracha'. The analyzed crosses were also characterized by a sufficiently high percentage of seedlings from fully formed seeds, which varied in the range from 56.0 to 63.3%, while all the indicators of seed germination were within the variation coefficient (13.1). The number of the obtained vigorous seedlings varied significantly (up to 83.6%).

The greatest number of vigorous seedlings was produced from combinations with the following varieties—'Alminskyi' (11.0%), 'Podarok Magaracha' (10.0%), and 'Pamyati Golodrigi' (9.0%). The values of the rated indicator for 'Granatovyi Magaracha', 'Antey Magarachsky', and 'Aurora Magaracha' varieties did not differ substantially from the mean in the sampling (5.5%). The smallest number of vigorous seedlings was observed in combinations with 'Pervenets Magaracha' (1.0%).

From the analysis of the obtained data, we can conclude that 'Podarok Magaracha', 'Alminskyi', and 'Pamyati Golodrigi' form a small number of seeds but can be reasonably used in crosses with the Crimean autochthonous grape varieties to produce vigorous seedlings. 'Pervenets Magaracha', on the contrary, does not transfer the powerful growth force to its progeny in these combinations.

To analyze the crossing capacity of the Crimean indigenous grape varieties, we carried out interspecific hybridization within *V. vinifera* species (**Table 6**).

As a result, the largest number of seeds was obtained from crossings with 'Kokur White' (1069 seeds), while the lowest number was obtained from crosses with 'Bastardo Magaraci' (112 seeds). The mean value of the resulting seeds made 649, with a slight variation coefficient in all the populations, which made 63. The standard deviation of the obtained hybrid seeds showed quite a considerable variation of 409. The percentage of fully formed seeds (seeds with a viable germ and well-developed endosperm) in variations made an average of 88.3%. 'Bastardo Magaraci' (99%) and 'Rkatsiteli' (77%) showed extreme values, which exceeded the variation coefficient as compared to the average value. As to the seed germination capacity, the hybrid varieties can be divided into two groups by reference to the mean value. The germination capacity of hybrid seeds obtained from crosses with 'Shabash', 'Kokur belyi', and 'Gevat Kara' was less than the average in variations of 54.9%; for

Male form, ♂	Hybrid seeds, pcs.	Fully formed seeds, %	Germinating capacity of hybrid seeds, %	Vigorous seedlings, %
'Kokur Belyi'	1069	88.0	47.5	12.5
'Gevat Kara'	1357	96.5	48.5	12.5
'Rubinovyi Magaracha'	232	95.5	59.5	10.0
'Bastardo Magaraci'	112	99.0	66.0	8.5
'Shabash'	549	95.0	46.0	7.0
'Rkatsiteli'	511	77.0	59.7	4.3
'Sauvignon vert'	681	80.3	56.0	3.7
'Chardonnay'	683	86.0	55.7	3.0
$\overline{x}$	649	88.3	54.9	7.7
σ	409	7.8	7.0	3.8
v	63	9.0	12.8	49.7

Table 6.
The effect of V. vinifera male forms on crossability.

the remaining species of *V. vinifera*, this indicator exceeded the mean value. The highest seed germination capacity was observed for 'Bastardo Magaraci' (66.0%).

Assessing male forms as to the transfer of the powerful growth force feature to the offspring, it may be noted that the largest number of vigorous seedlings was obtained in combinations with 'Kokur Belyi' and 'Gevat Kara' (12.5%). The mean values, 7.7% for variations, were exceeded by 'Rubinovyi Magaracha' (10.0%) and 'Bastardo Magaraci' (8.5%). The smallest percentage of the vigorous seedlings was observed for 'Chardonnay' (3.0%).

By comparing the impact on crossability of the Crimean autochthonous grape varieties of different male forms represented by varieties of *V. vinifera* species and those of interspecific origin, it is possible to ascertain some consistent patterns.

The observed variation coefficients of crossability indicators as to the formation of hybrid seeds obtained from intraspecific crosses within the species of *Vitis vinifera* are almost twice lower than in interspecific crosses. The established mean value for cross-combinations of the fully formed seeds differs insignificantly among intraspecific (within the species of *Vitis vinifera*) and interspecific crosses, with difference between them of only 1.9%. The established average percentage of seed germination capacity has determined a higher viability of hybrid seeds obtained as a result of cross-species hybridization.

The analysis of the studied cross-combinations allowed concluding that as a result of crossings within the species of *Vitis vinifera* the mean value of the obtained vigorous seedlings was almost 1.5 times lower than in interspecific crosses.

Thus, we were able to establish the genetically determined specificity of the crossability of the Crimean autochthonous grape varieties. By all crossability indicators of the Crimean autochthonous grape varieties, the variability was lower within the species of *Vitis vinifera* than in the interspecific hybridization. Using autochthonous variety 'Krona', as a female form, and interspecific variety 'Alminskyi', as male form, proved most efficient for obtaining vigorous progeny.

6. Modeling at plant breeding

In the middle of the twentieth century, there existed the so-called model of ideal variety of a grapevine—a concept that comprised the ultimate complex of the desired basic phenological and agrobiological characteristics in varieties that were to be cultivated on a vineyard [35]. But global and local climatic changes along with the change in biosphere conditions demanded creation of new cultivars, in particular in viticulture [36]. The global problem was not only to preserve the grapevine gene pool diversity but also to breed new grapevine cultivars that would correspond to modern conditions of biosphere [37].

Scientists in different countries discuss the problem and methods of creating a new genotype with given parameters [27], including the one on the basis of modeling [38]. A program for immunobreeding "Analogue" that is based on models of new grapevine cultivars was developed [39] and is being implemented at the "Magarach" Institute. The program aims to improve the efficiency of grape breeding and to achieve varieties, whose characteristics would be distinguished for genuine novelty. The breeding objective of the program is the development of new grape cultivars that would combine resistance to a variety of pests, diseases, and unfavorable environmental factors and would, at the same time, produce fruit of the quality comparable to the best samples of the existing international analogues. To achieve this, new baseline material is searched for, accumulated in collections, studied, and involved in generative hybridization.

Thanks to the existence of individual species of grapevine with peculiar traits of interest, and in pursuit of the increased heterogeneity of grapes under commercial cultivation, the grape breeding program "Analogue" aims at developing a new generation of grape cultivars based on the initial forms obtained from different centers: the European center (*Vitis vinifera L.*), the East Asian center (*Vitis amurensis* Rupr.), and the North American center (*Vitis riparia* Michx., *Vitis cinerea* Engelm., etc.).

The suggested concept of the models is based on 16 most desired selection traits that are used for the replenishment of the selection database, selection of the original forms for crossbreeding, evaluation of the selected hybrid forms, and multivariate analysis. Presently, the models of table grapevine cultivars are implemented in new varieties [3]. Using the developed models for the creation of new cultivars will help solve ecological problems in the vineyards, ensure high economic effectiveness of winegrowing, and receive grape cultivars for different purposes.

The basic signs defining phenotype and breeding value of genotypes (**Table 7**) are included in a model of a table grapevine variety. Varieties are distributed based on ripening time: super early (less than 105 days); very early (105–115 days); early (115–130 days); middle (130–140 days); and late (more than 140 days).

Since botanical, phenological, morphological, agrobiological, economic, physiological, and other signs are measured in different units and scales, it is necessary to formulate appraisal of trait manifestation in points. In other words, it is necessary to translate the quantitative and qualitative data into a uniform system. For this purpose the scope of a sign variation in the investigated set of cultivars was subdivide into 5 gradations with an interval of 2 (1, 3, 5, 7, and 9). The minimum value of the attribute is code 1, while the maximum is code 9. The presented model mapped genetic regularities of the maximum score in transgress hybrids. Formation of the maximum score is carried out based on quantitative (weight of a berry and that of a bunch) and qualitative (shape and color of a berry) attributes that are associated with the ripening times.

The conducted hybridological analysis and mathematical and statistical processing of the experimental data highlighted the most valuable combinations of crosses on complex inheritance of phenotypic traits of elegance with a bias toward early ripening (**Table 8**).

Combinations of 'Flora' × 'Rishel'e' (43.9%), 'Present to Zaporozhye' × 'Rishel'e' (63.9%), and 'Flora' × 'Find of Mariupol' (100%) have the highest breeding value in terms of the ripening period. The degree of the

Ripening time	Super early	Very early	Early	Middle	Late
	9	7	5	3	1
Berry weight	5	7	9	7	5
Bunch weight	5	7	9	7	5
Berry shape	3	5	7	9	3
Berry coloration	9	7	5	3	1

Legend. Sign expression in points: Berry weight: 5 points, 6 g; 7 points, 8 g; 9 points, more than 10 g; Bunch weight: 5 points, 500 g; 7 points, 800 g; 8 points, 1000 g; 9 points, more than 1200 g; Berry shape: 3 points, rounded; 5 points, egg-shaped; 7 points, cylindrical; 9 points, elongated-elliptic; Berry coloration: 1 point, blue-black; 3 points, violet; 5 points, red; 7 points, pink; 9 points, yellow-green.

Table 7.
Model of a table grapevine cultivar according to the expression of phenotype traits.

Crossing combination	Ripening time	Berry weight	Bunch weight	Berry shape	Berry coloration
'Flora' × 'Rishel'e'	43.9	33.7	10.3	43.9	39.8
'Flora' × 'Cardinal'	12.4	0.0	10.3	0.0	45.4
'Flora' × 'Find of Mariupol'	100	20.0	20.0	40.0	100.0
'Talisman' × 'Cardinal'	9.1	100.0	50.0	0.0	40.9
'Talisman' × 'Kodryanka'	11.3	100.0	51.0	0.0	30.2
'Present to Zaporozhye' × 'Rishel'e'	63.9	84.8	20.0	9.8	25.6
'Present to Zaporozhye' × 'Cardinal'	10.9	84.6	20.0	0.0	32.0
'Flamingo' × 'Rishel'e'	11.5	8.0	8.0	0.0	23.0
'Flamingo' × 'Arcadia'	9.8	75.0	25.0	0.0	38.2

Table 8.
Breeding value of grapevine populations (%).

hypothetical heterosis of the hybrid offspring as to this trait ranged from 0.6 to 66.7% (**Table 9**).

Populations that include 'Talisman' variety show the highest breeding value of the inheritance of a big berry, reaching up to 100%. The high breeding value is due to the biological capacity of the 'Talisman' variety to generate a very big berry, the weight of which reaches up to 24 g, and a high degree of transmission of this trait to its progeny in populations.

Hereby, the overall combining ability of the 'Talisman' variety influence on the increase of the berry weight in progeny as compared to the average indices of the original forms happens with a fairly low severity of 1.25%. Reverse causality between heterosis and breeding value is observed in combination of 'Flora' × 'Cardinal'. In this case there is a zero degree of inheritance of giant berries; however, there is a significant (63%) superiority of the hybrid progeny over the average index of a berry weight, characteristic of both parents.

The breeding value of the inheritance of a bunch weight in represented populations ranged from 8 to 51%. It should be noted that the highest conjugate of

Crossing combination	Ripening time	Berry weight	Bunch weight	Berry shape	Berry coloration
'Flora' × 'Rishel'e'	5.7	8.3	1.9	−1.9	−2.3
'Flora' × 'Cardinal'	1.4	63.0	6.7	35.0	0.2
'Flora' × 'Find of Mariupol'	66.7	9.3	5.7	−3.8	0.0
'Talisman' × 'Cardinal'	0.6	1.3	8.8	−4.5	1.3
'Talisman' × 'Kodryanka'	8.9	1.2	12.1	4.2	−12.8
'Present to Zaporozhye' × 'Rishel'e'	34.7	3.0	−5.7	−17.9	−23.5
'Present to Zaporozhye' × 'Cardinal'	5.1	5.5	0.7	−7.3	−18.7
'Flamingo' × 'Rishel'e'	9.8	7.3	−14.4	−3.1	−7.5
'Flamingo' × 'Arcadia'	14.5	−3.3	−4.3	2.7	0.8

Table 9.
Hypothetical heterosis in grapevine populations (%).

the studied traits is observed between the weight of a berry and the weight of a bunch and a distinctive relationship is observed in crosses with 'Talisman' variety. Populations, in which this variety participates, produce about 50% progeny forms with a very big bunch.

In this study preference was given to the elongated oval-shaped berries. However, very early-ripening cultivars mainly produce a rounded berry. Apparently, this feature is associated with the biological specificity of the berry formation and the origin of table grapes *Vitis vinifera orientalis* Negr. that is related to soil and climatic conditions of their origin. Many varieties of eastern ecogeographical group, 'Husaine Kelim Barmak', 'Shami Abiad', 'Fakhri', and others, have an elongated oval or cylindrical berry shape and are characterized by the average ripening time.

The main objective of a plant is to generate progeny and in the case of grapes to produce seeds. It is known that very early-ripening cultivars have zero seed germination due to the insufficient formation of a normally developed embryo and endosperm, while the seeds of the average ripening period cultivars have the highest viability. It is possible that the genetic correlation between the shape and weight of berries and the ripening period is due to the biological dependence of the length of the vegetation period and formation of a sufficient amount of biochemical compounds. However, the studied hybrid progeny has a high degree of heterozygosity. Initial forms of interspecific origin represent at least 5 generations and have more than 50 ancestors. The oval shape of a berry is intermediate between the round and cylindrical forms. Therefore, in the process of breeding early-ripening cultivars with oval berry shape, there is a sufficiently high degree of probability to obtain hybrids with this berry shape. The most valuable populations in this case were obtained by crossing with 'Flora' cultivar, specifically 40–43.9%, in the genotype of which 'Khusaine Belyi' genes are present.

Berry color is a sign closely related to the preservation of grapes. The higher the percentage of anthocyans (phenolic compounds) in the skin of the berries, the longer is the period during which they can be preserved. The economic value of this feature increases with the elongation of the vegetation period. Very early-ripening varieties do not require long-term storage, as fresh grapes are sold without storage and are in high demand. With the increase in the vegetation period, however, the potential yield of varieties increases, along with their competitiveness. Thus, it develops the need for a short-term storage of early varieties, smoothly developing into the long-term storage of late varieties. Therefore, the model of a table variety displays the optimal berry coloration resulting from the content of anthocyans (phenol compounds) and aroma compounds [30], associated with the ripening period and storage duration.

According to the existing genetically determined pattern, the dark color of a berry dominates over a lighter color. Using this pattern has created a population with sufficiently high breeding value of 23–100% for the given trait. However, the hypothetical heterosis in populations with colored varieties—'Rishel'e', 'Kodryanka', and 'Cardinal'—has a negative value, underscoring the validity of color patterns of domination.

In the assessment, populations were grouped into separate complexes by the aggregate of model characteristics. Experimental data on the breeding value and heterosis were obtained (**Figure 3**). Using this approach in the analysis, it is possible to consider the overall breeding value of each population and reveal the general and specific combining ability of the initial forms. As a result, the most effective crossbreeding combinations were identified with the purpose of obtaining model for early-ripening table cultivars. The following combinations have the biggest breeding value: 'Present to Zaporozhye' × 'Rishel'e', 'Talisman' × 'Cardinal', and 'Flora' × 'Find of Mariupol'. The overall positive combining ability is characteristic

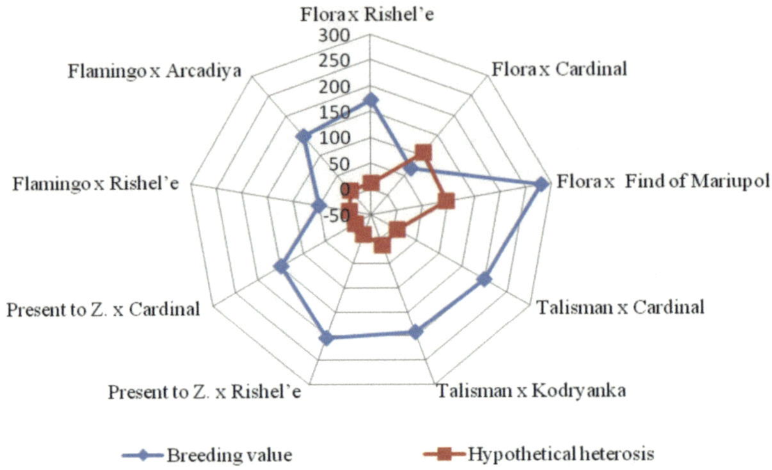

Figure 3.
Degree of expressiveness of model signs in grapevine populations.

of the original forms of 'Flora' and 'Talisman', negative —'Present to Zaporozhye'. The positive specific combining ability was observed in 'Flamingo' variety in crosses with 'Arcadia' variety. By the total common indicators, of special interest is the combination of 'Flora' × 'Find of Mariupol'.

The analysis of the inheritance of the aggregate model features allowed creation of new table cultivars of very early-ripening period with big berries and bunches: 'Solnechnaya grozd' ('Flora' × 'Find of Mariupol'), 'Liviya' ('Flamingo' × 'Arcadia'), and 'Academician Avidzba' ('Present to Zaporozhye' × 'Rishel'e').

The conducted research established global and local climate change and the associated change in the biotic stress factors of the biosphere. This problem results in a need to cultivate new grape cultivars that are environmentally plastic and resistant to the biotic and abiotic factors of the biosphere. The targeted creation of new grape cultivars based on formulation of models and their subsequent implementation is perceived as the most successful approach.

The hybridological analysis of the newly created grape hybrid gene pool revealed the most effective combinations of crosses for early-ripening cultivars with a big berry. Also, the general and specific combining ability for certain initial forms that appear in the progeny was identified. Overall, the model practice allowed obtaining new early-ripening grape cultivars with big berries that are being successfully cultivated in the vineyards under different environmental conditions.

7. Aspects of the general and particular in the formation of programs of fundamental scientific research in horticulture

The modern conditions of the biosphere and the cultivation of agricultural crops, which include horticulture and grapes in particular, determine the need for the formation of scientific research programs, the implementation of which would improve the efficiency of agricultural production, in particular of these crops. The analysis of the complex of scientific researches makes it possible to note that genetic selection and physiological-biotechnological priorities can be considered as priority directions and effective implementation of specific scientific developments is possible only after taking into account the existing environmental factors of biotic and abiotic nature.

It is really possible to make an intellectual breakthrough in solving the problems of improving the methodology and scientific foundations of breeding genetic research, as well as increasing the efficiency of agricultural production only on the basis of the development and implementation of national programs. The leading role in the development of these programs should be carried out by the national scientific institutions of the country, to which the All-Russian National Research Institute of Viticulture and Winemaking "Magarach" belongs in Russia. At the present time, in the "Magarach" Institute has developed and is implementing a genetic selection program for the breeding of new grape varieties "Analogue." This program provides for the modeling of new grape genotypes based on the developed models. In turn, the formation of models of new varieties of grapes is based on established patterns of manifestation and inheritance of selectable features selected for each model according to breeding value, crossability, hypothetical heterosis, and other genetic selection patterns.

Participation of scientists of the Magarach Institute in international European scientific projects, including EU projects, gives grounds to say that the research in the institute corresponds to the modern world level of development of both fundamental biological and agronomic sciences.

Among a number of prerequisites necessary for the stable and efficient functioning of any crop production, a special place belongs to the sustainability of its development. This concept includes a whole range of organizational, economic, social, and environmental assessments. The most important of them for agricultural production is the requirement to adequately meet the needs of living people, without depriving this opportunity for future generations, the need to harmonize the way of life with the ecological capabilities of the region, and the introduction of certain restrictions on the exploitation of natural resources due to the ability of the biosphere to cope with the consequences of human activities. This is consistent with the international convention on the ecological status of the biosphere on the planet.

Sustainable development is the main condition for the effective functioning of agricultural production. At the heart of sustainable development of agricultural production are several main factors. One of them is environmental conditions, the requirements for which are presented by living organisms—soil, plants, animals, birds, and insects—which are diverse due to the biological characteristics and diversity of living organisms themselves. The yield of crops and, as a consequence, the economic efficiency of production are determined by the extent to which the living conditions correspond to the requirements of the biology of living organisms. No other circumstances, the provision of machinery, pesticides, and fertilizers, although important, cannot replace this requirement. From this basic condition follows the main task for agriculture, which consists in creating in each soil climatic zone such biocenoses, whose biological requirements would be most adequate to the conditions of their functioning. Only in this case, it is possible to count on their optimum productivity. The main functional unit of the agricultural biocenosis is a variety. Naturally, in these conditions, it is necessary to pay priority attention to the use of biological reserves of the plants themselves of agricultural crops. In other words, it is necessary to go as far as possible toward the creation and use of new selection varieties of grapes that meet modern requirements of agricultural technology and the conditions of the biosphere to the maximum.

This is also confirmed by the fact that in the complex of biological sciences that solve the tasks of increasing the productivity of agricultural production, the most solid positions belong to genetics. At the same time against the background of changing climatic conditions and an increase in the epiphytotic load of phytopathogens in agrocenoses, it becomes most expedient to use varieties of a new breeding generation and clones of traditional varieties resistant to biotic and abiotic factors.

This direction of selection is characteristic for scientific research in Russia and in all developed countries of Europe and the world.

Increasing the efficiency of agricultural production, including horticulture crops and grapes, is really possible only on the basis of the large-scale use of modern scientific and technical developments. The increase in the level of scientific and technological achievements and their use will enable the most fully to realize the high biological potential of the plant and thereby increase the supply of food to the population. At the same time, the priority areas of research are the creation of a new generation of winter-hardy, immune, and productive varieties providing high-quality agricultural products, which are the basis of adaptive horticulture growing.

The criteria for breeding new generations of varieties of horticulture crops can be based on the same principles used in breeding programs adopted for individual crops, particularly for grapes. Based on knowledge of the existence of separate centers of origin of cultures, the evolutionary variability of plants, and the coevolution of plants and pathogens, it is possible to select the necessary initial forms for generative hybridization, which currently remains the main method in breeding. It is also possible to use the laws of genetically determined patterns of crossing and inheritance of signs and characteristics in individual crops when selecting other horticulture crops. Improvement of the methodology of breeding for the purpose of breeding a variety with the necessary parameters is possible, in particular, based on the development and implementation of models of new varieties, which take into account the knowledge of general and private genetics.

Author details

Vladimir Volynkin*, Alla Polulyakh, Svetlana Levchenko, Irina Vasylyk and Vladimir Likhovskoy
All-Russian National Research Institute of Viticulture and Winemaking "Magarach", Yalta, Russian Federation

*Address all correspondence to: volynkin@ukr.net

References

[1] OIV. OIV Descriptor List for Grape Varieties and Vitis Species. 2nd ed. Paris: O.I.V. (Off. Int. Vigne Vin); 2001. p. 227

[2] OIV. Codes des caracteres descriptifs des varietes et especes de Vitis. 2009. Available at: http://www.oiv.int/fr/ and http://www.oiv.int/oiv/info/frplubicationoiv#listdesc

[3] Volynkin V, Polulyah A, Klimenko V, Lihovskoi V, Oleinikov N, Levchenko S, et al. Breeding for Ukrainian table grape varictics. Vitis. 2015;**54**:157-158

[4] DeLattin G. On the genetics of grapevines. Present results of the factor analysis of Vitis. Vitis. 1957;**1**:1-8

[5] Peakall R, Smouse PE. Genalex 6: Genetic analysis in excel. Population genetic software for teaching and research. Molecular Ecology Notes. 2006;**6**:288-295

[6] Tukey JW. Data analysis, computation and mathematics. Quarterly of Applied Mathematics. 1972;**30**:51-65

[7] Seehausen O. Hybridization and adaptive radiation. Trends in Ecology and Evolution. 2004;**19**:198-207

[8] Cabezas JA, Cervera MT, Ruiz-Garcia L, Carreño J, Martínez-Zapater JM. A genetic analysis of seed and berry weight in grapevine. Genome. 2006;**49**: 1572-1585

[9] Schippers PA. Composition changes in the nutrient solution during the growing of plant in recirculating nutrient culture. Acta Horticulturae. 1980;(98):103-118. DOI: 10.17660/ActaHortic.1980.98.9

[10] Peralbo A, Flores F, López-Medina J. Recirculating nutrient solution in strawberry. Acta Horticulturae. 2005;(697):101-106. DOI: 10.17660/ActaHortic.2005.697.11

[11] Golodriga PY. The methodology for the selection and evaluation of hybrid grape seedlings. In: Proceedings of the All-Russian Viticulture and Enology Scientific and Research Institute “Magarach”, V. XII. 1963. pp. 25-35. (in Russian)

[12] Melkonyan M, Volinkin VA. Methods of Ampelographic Description and Assessment of Agrobiological Evaluation of Grapes. Yalta: Institute of Grape and Wine “Magarach”; 2002. pp. 27. (in Russian)

[13] Klimenko VP. Guidelines for Quantitative Genetics of Grapes. Yalta: Institute of Vine and Wine “Magarach”; 1998. pp. 24. (in Russian)

[14] FAO. The Future of Food and Agriculture—Trends and Challenges. Rome: Food Agric. Org. United Nations (FAO); 2017. 34 p

[15] Eiras-Dias JEJ. Status of the Vitis national collection in Portugal. In: Report of a Working Group on Vitis. Rome, Italy: Bioversity International; 2008. pp. 93-94

[16] Lacombe T. Status of the French Vitis National Collection. In: Report of a Working Group on Vitis. Rome, Italy: Bioversity International; 2008. pp. 73-74

[17] Maghradze D, Maletic E, Maul E, Faltus M, Failla O. Field genebank standards for rapevines. Vitis. 2015;**54**: 273-279

[18] Pelengic R, Koruza B. Slovenia grapevine germplasm. Acta Agriculturae Slovenica. 2012;**99**(3):429-432

[19] Li SH, Archbold D, London J. Collection, conservation, evaluation and

utilization of *Vitis amurensis* germplasm resources in China. Acta Horticulturae. 2015;(1082):79-86

[20] Dettweiler E. Genetic resources—Gene banks. Vitis. 1990;**29**:57-59

[21] Volynkin V, Polulyakh A, Chizhova A, Roshka N. Ukraine: Native varieties of grapevine. Caucasus and Northern Black Sea region Ampelography. Vitis. 2012:405-473

[22] Dettweiler E, This P, Eibach R. The European network for grapevine genetic resources conservation and characterization. In: Materials of XXV Congress mondial de la vigneet du Vin. France, Paris. 2004. pp. 1-10

[23] Soltis DE, Morris AB, McLachlan JS, Manos PS, Soltis PE. Comparative phylogeography of unglaciated eastern North America. Molecular Ecology. 2006;**15**:4261-4293

[24] Pe'ros J-P, Berger G, Portemont A, Boursiquot J-M, Lacombe T. Genetic variation and biogeography of the disjunct Vitissubg. Vitis (Vitaceae). Journal of Biogeography. 2011;**38**: 471-486

[25] Barrett HC, Carmer SG, Rhodes AM. A taximetric study of interspecific variation in Vitis. Vitis. 1969;**8**:177-187

[26] Soejima A, Wen J. Phylogenetic analysis of the grape family (Vitaceae) based on three chloroplast markers. American Journal of Botany. 2006;**93**: 278-287

[27] Töpfer R, Hausmann L, Eibach R. Molecular breeding. In: Adam-Blondon A-F, Martinez-Zapater J-M, Kole C, editors. Genetics, Genomics and Breeding of Grapes. Enfield, New Hampshire, USA: Science Publishers; 2011. pp. 160-185

[28] Myers N, Mittermeier RA, Mittermeier CG, da Fonseca GAB, Kent J. Biodiversity hotspots for conservation priorities. Nature. 2000; **403**:853-858

[29] Ricklefs RE. Historical and ecological dimensions of global patterns in plant diversity. Biologishe Skrifter (Royal Danish Academy of Sciences and Letters). 2005;**55**:583-603

[30] Levchenko SV, Volynkin VA. The aroma-forming substances of hybrid seedlings of grape. Acta Horticulturae. 2014;(1046):523-530

[31] Maghradze D, Turok J. Project of bioversity International on conservation and sustainable use of grapevine genetic resources in the Caucasus and northern Black Sea region: Activities and results. In: Maghradze D, Rustion L, Turok J, Scienza A, Faila O, editors. Caucasus and Northern Black Sea Region Ampelography. Vitis – Journal of Grapevine Research. 2012. pp. 17-25

[32] Forni G. The origin of "Old World" viticulture. In: Maghradze D, Rustion L, Turok J, Scienza A, Faila O, editors. Caucasus and Northern Black Sea Region Ampelography. Vitis – Journal of Grapevine Research. 2012. pp. 27-38

[33] De Lorenzis G, Simone Di Lorenzo GB, Failla O, Musayev MK, Salimov V, Maghradze D, et al. Study of genetic diversity in *V. vinifera* subs. sylvestris in Azerbaijan and Georgia and relationship with species of the cultivated compartment. Act Horticulturae. 2015;(1074):49-53

[34] Volynkin V, Polulyahk A. Origin of grapevine varieties in Crimea and *Vitis vinifera subsp*. sylvestris classification. Vitis. 2015;**54**(Special Issue):227-228

[35] Golodriga PY, Troshin LP. Biology and technology program for the creation of highly complex-resistant varieties of grapes. In: Golodriga PY, Guzun NI, Nedov PN, editors. Prospects of Breeding and Genetics of Grapes on the

Immune System. Kiev, USSR: Naukova Dumka; 1978. pp. 259-264 (in Russian)

[36] Jones GV, White MA, Cooper OR, Storchmann K. Climate change and global wine quality. Climate Change. 2005;**73**:319-343

[37] Duchene E, Huard F, Dumas V, Schneider C, Merdinoglu D. The challenge of adapting grapevine varieties to climate change. Climate Research. 2010;**41**:193-204

[38] Chaïb J, Torregrosa L, Mackenzie D, Corena P, Bouquet A, Thomas MR. The grape microvine—A model system for rapid forward and reverse genetics of grapevines. The Plant Journal. 2010;**62**: 1083-1092

[39] Volynkin VA, Klimenko VP, Oleinikov NP. Grape selection for immunity based on the models of grape varieties. In: VI International Symposium On Grape breeding, Yalta, Crimea, 4–10 September. 1994, 1994. pp. 72-73

Section 3

Epidemiology and Management of Pests and Diseases

Chapter 6

Epidemiology and Management of South American Leaf Blight on Rubber in Brazil

Edson Luiz Furtado, Willian Bucker Moraes, Waldir Cintra de Jesus Junior, Breno Benvindo dos Anjos and Lilianne Gomes da Silva

Abstract

The rubber tree (*Hevea* spp.) is one of the main forest crops in tropical regions due commercialization of natural rubber. Brazil currently imports most rubber that is consumed. According to the International Rubber Study Group, for an annual consumption of 350,000 tons in Brazil, 135,000 tons were produced, whereas 215,000 tons were imported. This failure of rubber cultivation in Brazil is primarily due to South American leaf blight (SALB), a disease caused by the fungus *Microcyclus ulei* (P. Henn. v. Arx.). The fungus is present in all Brazilian rubber-producing regions and attacks young leaflets, causing abscission and, ultimately, death of the tree. This disease occurs in almost all areas of rubber tree plantations in Central and South America. Strategies used to manage SALB are based on the use of fungicides in nurseries and young plantations and the use of resistant clones; on phenological aspects, taking into account the leaf shedding patterns of adult rubber trees, which in certain environments provide defense in addition to resistance; and on climatic factors that are favorable or unfavorable to epidemic development. The aim of this chapter was to describe all aspects related to the epidemiology and management of leaf blight in Brazil.

Keywords: *Microcyclus ulei*, *Hevea brasiliensis*, control, evasion, damage and losses

1. Introduction

From 1914 to the 1970s, South American leaf blight (SALB) was considered one of the major causes of failure of commercial rubber cultivation in South America [1]. Today, management measures are available to ensure a minimum risk of epidemics to rubber tree crops in several regions of Brazil.

Among such measures is planting of resistant plant materials associated with evasion (choice for areas favorable to rubber cultivation and unfavorable to the pathogen). These actions have reduced injuries and losses caused by this disease and, consequently, favored large-scale production in Brazil and in other countries that aim at self-sufficient production of rubber, a strategic raw material.

This disease was described at the early twentieth century in leaves collected from native rubber trees in the surroundings of Belém, Pará State (PA); symptoms were few, not causing defoliation or other injuries to the plants since susceptible rubber trees grow naturally at a low density in forests, 3–4 trees per ha.

However, the devastating potential of this disease was detected in the first attempts to domesticate this species and establish commercial plantations in the Guianas and Brazil (Ford Crops) at the beginning of the century. Such failures are documented in the literature but were not taken into account by the Brazilian authorities responsible for the sector's policy and fiscal incentives, as well as by the former support agencies, which made some historical mistakes. One of these mistakes was the financial support by the Brazilian government to rubber cultivation at the humid Amazon region, with goals of planting rubber trees in 250,000 ha from 1970 to 1985 (PROBOR I, II, and III). Of these, only 120,000 ha were planted, of which 38,000 ha were decimated by the disease and the remaining had very low yield, discouraging new cultivation and investments in the sector [2]. Another mistake was the choice for hybrid clones of *Hevea benthamiana* and *H. brasiliensis*, which present low yield, do not have uniform leaf shedding, and do not allow breaking the pathogen's cycle.

At the end of this period, when the financial aid to the sector had finished, pressure by other Brazilian states allowed cultivation in other areas of the country, such as the central region, part of the central-west region and the southeast region, where rubber trees were grown exuberant and productive, free of leaf blight epidemics. Nowadays, these are considered areas of high productivity due to disease evasion (erroneously cited in the literature as "scape areas"), showing the way to high productivity and national self-sufficiency of such strategic raw material. Details will be addressed in this chapter.

2. Etiology of the disease

South American leaf blight, or leaf blight, is caused by the fungus Ascomycota *Microcyclus ulei* (P. Henn) v. Arx, which is found throughout cultivation areas in the American continent and is particular to the genus *Hevea*; it was already detected in six rubber tree species [3].

This fungus has two types of infective spores in its life cycle, according to the type of reproduction: conidiospores (asexual reproduction or anamorphic phase), the anamorphic reclassification of which was recently described in the literature, e.g., *Pseudocercospora ulei* [4], with cycle varying between 6 and 10 days, depending on the clone, and ascospores (sexual reproduction or teleomorphic phase), with cycle varying from 100 to 150 days (**Figure 1**).

Conidia are responsible for spreading the pathogen and causing epidemics due to their large number and rapid cycle. Ascospores take longer time and are responsible for the primary inoculum, minimally contributing to the epidemics; they can affect young plants and clonal gardens or species and hybrids that do not regularly change leaves (*H. benthamiana*). These sexual spores are produced in a small quantity and remain protected inside special structures on the leaves (pseudothecia) for several months, even in fallen leaflets, being progressively released to the air.

In nurseries at Paraíba Valley and South Bahia, weather conditions favorable to epidemics included relative humidity superior to 95% during 10 consecutive hours for 12 days a month. The disease affects specially leaflets but can also reach petioles, new branches, and even fruits [6].

Once spores come in contact with leaflets in the susceptible stage, initial symptoms appear on the inner face of the leaf as small rounded necrotic spots, under which dark green conidial sporulation of velvety aspect emerge in most clones. Under high humidity conditions, lesions grow occupying great part of the leaf limb and causing defoliation.

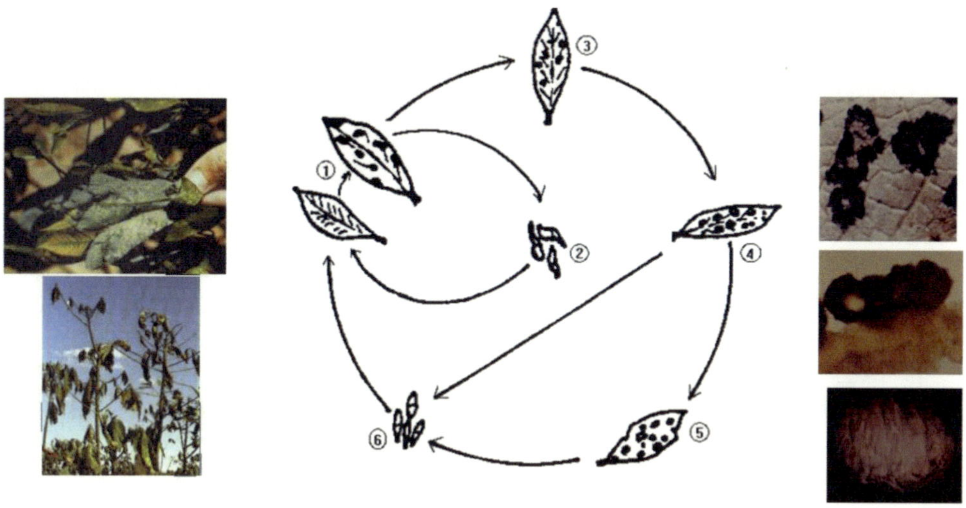

Figure 1.
Phases of Microcyclus ulei cycle, modified from [5]. (1) Wet period. Infoliation; (2) Anamorphic phase. Conidial sporulation. Leaflet shedding; (3) Teleomorphic phase. Stromata formation; (4) Mature stromata; (5) Dry period. All plants changing the leaves; ascospores released; and (6) Ascospores released by leaflets dropped. New infections occurred and beginning a new cycle.

In the remaining infected leaflets, the sexual phase develops (spermatogonia, asci, and *Ascospores*); ascospores consist in formations showing diameter of several millimeters that become massive and rough to the touch, like sandpaper (stroma). Such symptoms prevail in mature leaflets until their natural fall.

3. South American leaf blight in native rubber trees

Native rubber trees are intertwined with the Amazon forest, where they are inserted, occupying several Brazilian states and Amazon countries of South America, in which the 11 known species of *Hevea* cohabit, showing varied size (shrub to tree), variable phenology (see text below), and variable resistance to the causal agent of leaf blight. Among an enormous diversity of plant species, rubber trees are rare, i.e., they have extremely low density (2–3 plants per hectare). Such rarity and diversity protect them from herbivory and from the attack by pest insects and diseases, such as leaf blight. Based on studies conducted with native rubber trees in Acre State, Furtado [7] verified the percentage of diseased fallen leaflets per plant (incidence). In addition, diseased leaflets were evaluated for the percentage of injured leaf area (severity), according to the diagrammatic scale developed by Chee [8] and modified by Gasparotto [1]. Data were collected from native rubber trees or settings (areas of ~400 ha) containing three lots of 150 rubber trees, which were productive and aged more than 100 years ("rubber tree roads"), located at Chico Mendes Extractive Reserve, in the rubber tree plantations: São Pedro, Dois Irmãos, Nazaré, and Floresta, and at Caquetá Extractive Settlement, totaling 11 native rubber tree roads. Sampling included one out of every two individuals. Fifty trees per "road" of native and planted rubber trees were covered (**Table 1**).

There were no epidemic levels in any of the rubber tree plantations; only a greater leaf blight incidence was noticed for rubber tree plantations in Nazaré and São Pedro, evidencing a balanced situation between the native rubber trees and the fungus.

Rubber plantation	Rubber tapper's name	*Microcyclus* Incidence	*Ulei* sev.
São Pedro			
Bom levar II	Raimundo Carlos	12	0.6
Vai quem Quer	Dalvo F. da Silva	12	0.1
Morada Nova	Antonio M. E. de Oliveira	9	0.15
Floresta			
Bela vista I	Domingos F. da Conceição	0	0
Maloquinha	Francisco das C. F. Marcelino	0	0
Bom Principio	Raimundo N. P. da Silva	8	0.3
Taripu (Enrascado)	Manoel da S. Oliveira	0	0
Nazaré			
Rio Branco II	Guilherme Q. de Oliveira	12	0.4
Caqueta			
Limoeiro	Ladislau do Nascimento	0	0
São Pedro	Acelino do Nascimento	0	0
Feijão Duro	Osmã A. da Silva	0	0

Table 1.
Intensity of leaf blight symptoms obtained for native rubber trees at different settings/plantations. Acre Valley [7].

4. Epidemiology of South American leaf blight

Learning the structure and the behavior of leaf blight is essential for the rational management of this disease. Thus, the dynamics of the pathogen's ascospores and conidia must be quantified, as well as the phenology of the rubber tree, the intensity, and progress of the disease, and the effects of altitude and leaf density on the production and growth of the rubber tree and leaf wetness period significantly influence the *Hevea* sp. X *M. ulei* pathosystem.

Infection by *Microcyclus ulei* occurs in young leaves of rubber trees at the optimum temperature for the disease, around 24°C. However, temperatures between 20 and 28°C are favorable to the disease, evidencing high number of foliar lesions [1].

The pathogen *M. ulei* can infect rubber trees at 16°C and the disease can evolve normally at 24°C, showing that the temperature of 16°C does not prevent the pathogen from penetrating the leaves of rubber trees but makes colonization slow or paralyzed [9].

SALB can occur in rubber trees after 6 hours of leaf wetness at 24°C since it highly depends on the aggressiveness of isolates and on whether the clone is susceptible or resistant. Localities subject to dew, fog, or light rain for prolonged periods present conditions extremely favorable to the development of leaf blight [1]. The longer the low temperature period, the minor the disease severity.

The deciduous habit of the rubber tree is considered important since it reduces the initial inoculum, located in older leaves, and standardizes the buddings, which are very important in scape areas, where refoliation coincides with the shortest leaf wetness period, unfavorable to the development of the pathogen [10].

Honorato [11] quantified the following variables: dynamics of the pathogen's ascospores and conidia; phenology of rubber trees; disease progress; and effects of altitude, disease severity, and leaf density on the production and growth of rubber trees. Experiments were conducted in rubber tree plantations at Igrapiúna, Bahia, considering three topographic conditions (top, slope, and lowland). The author detected ascospores and conidia of the fungus over the whole experimental period, but ascospore concentration was higher in the nocturnal period, while that of conidia was greater in the diurnal period. The effect of climate variables on conidial release was greater at lowland. At the top, leaf density of the rubber tree was higher and the disease severity was lower. Climate variables had a greater effect on disease severity at lowland, where the number of hours with leaf wetness was higher and relative humidity was minimal. A mean reduction of 47.7% in rubber production was detected at lowland, as well as mean severity of 15.0% and mean reduction in leaf density of 50.1%. There is evidence to propose changes in the pathogen's life cycle since ascospore and conidial production occurs all over the year in the field, under favorable environmental conditions. According to the obtained results, the author concluded that the effects of environmental variables on the disease are more evident under lowland conditions, where leaf blight, in particular, reduces the rubber production and the rubber tree growth. Under such conditions, the disease management measures should be intensified, including the planting of clones with high horizontal resistance.

An interesting phenomenon that has become frequent in the last years is the so-called South Atlantic Convergence Zone (SACZ), defined as a region of extensive cloud bands from the Amazon, Central-West, and Southeast region [12]. This climatological characteristic is associated with rainfalls that sometimes are strong, sometimes are moderate, and sometimes are intermittent, persisting for a minimum of 4 days, which can cause great disorders like floods, landslides, and overflows.

Climatologically, SACZ system is responsible for the considerable amount of summer rainfall among the Central-West, Southeast, and part of North and Northeast regions, causing humidity accumulation during the summer, which must have favored such leaf blight epidemics at the beginning of the year in São Paulo, Mato Grosso do Sul, Mato Grosso, and Goiás States. On the other hand, the absence of this system causes drastic reduction in rainfall in these regions, consequently leading to losses in agricultural production and risk of water and energy rationing [13].

As SACZ phenomenon is common in Brazil, new leaf blight epidemics are expected when a highly humid period coincides with the presence of new leaves in the rubber tree, since the pathogen is common, at low intensity, even in "scape" zones. The use of productive clones resistant to leaf blight is welcome, even for "scape" zones [13].

5. Strategies for the management of South American leaf blight in rubber tree plantations

The strategies used in the management of this disease in Brazil follow the proposals published [2, 3], based on the rubber tree-climate-*M. ulei* interaction, i.e., the phenological features of the used clones (which must have a uniform deciduous habit), on the resistance to leaf blight, and on the climate characteristics favorable or not to epidemics, in each Brazilian region, modified from Ortolani [14], shown in **Table 2**.

5.1 Phenological behavior of the rubber tree

The deciduous habit of the rubber tree is one of its most important phenological features for allowing yearly leaf renovation. Considering this character, the major species of rubber tree and their hybrids planted in Brazil present a highly important variation:

a. *Hevea brasiliensis* has the largest number of hybrids with uniform and regular deciduous habit, and their leaf shedding occurs from August to September, which corresponds to the dry period. Thus, diseased leaflets are naturally threshed and healthy growth occurs after hibernation and new budding.

I – Amazon Region:

AM1 – Marginal area, of constant super humidity and outbreaks of the disease. Annual water deficit (Da) = 0 mm; mean relative humidity of the driest month (RHs) > 85% and real evapotranspiration (RE) > 900 mm. Western Amazonas.

AM2 - Marginal area, high humidity and outbreaks. Da 0-100 mm, RHs 75-85% and RE > 900 mm. Central Amazonas.

AM3 - Marginal and preferential area with restrictions. Moderate to high disease incidence. Obligate phytosanitary control although there is a variable dry season. Da 100-200 mm, RHs 65-80% and RE > 900 mm. Eastern Amazonas.

AM4 - Preferential area with restrictions. Low incidence of *M. ulei*. Caution is required for the establishment of rubber tree plantation due to high seasonal water deficit. Da 200-300 mm, RHs 65-80%. It covers the area of transition between Central Brazil and dense Forest.

II – Non-Amazon Regions:

A - Preferential area with satisfactory thermal and water conditions and minimal risk of disease incidence. Da 0-200 mm; RHs 55-70% and RE > 900 mm.

A1 - Preferential area with restrictions. Low disease incidence. Caution is required for the establishment of rubber tree plantation due to seasonal water deficit (Da 200-300 mm).

B - Marginal area with super humid conditions. Moderate to high disease incidence. Obligate phytosanitary control. Da=0 mm, RHs > 80%, mean temperature of the coldest month (Tf) > 20°C (e.g., South Coast of Bahia).

B1 - Marginal area with super humid conditions. Moderate to high disease incidence in clonal gardens, nurseries and new plantations, or adult plantations with cultivars that do not adequately shed their leaves (hybrids of *H. benthamiana*). It differs from the previous region for presenting Tf < 20°C or longer dry period in leaf shedding (e.g., Ribeira Valley Region).

C, D and E – Marginal to inapt areas due to thermal or water limitations.

Table 2.
Brazilian climate zoning for the rubber tree, aimed at controlling South American leaf blight, modified from [14].

b. *Hevea benthamiana* and its natural hybrids, or with *H. brasiliensis*, present irregular leaf shedding, partially changing their crown every year in the dry period. Part of the diseased leaves remain in the crown and will serve as initial inoculum for new leaflets during budding.

c. *Hevea pauciflora* and its hybrids do not have a defined period for leaf shedding, being intertwined with perennial species. However, they have high resistance to leaf blight.

5.2 Resistance of rubber trees to South American leaf blight

In Brazil, breeding with rubber trees started in 1937 after outbreaks caused by the fungus *M. ulei* in crops established by Ford Company in the fields of Fordlândia in 1928 and in Belterra in 1932, both at low Amazonas, Pará State. Currently, the species of greatest interest for breeding are: (1) *H. brasiliensis*—higher productive capacity and genetic variability for resistance to *M. ulei*; (2) *H. benthamiana*—resistance to *M. ulei* and genetic variability for rubber production; (3) *H. pauciflora*—certain immunity to *M. ulei*; and (4) *H. camargoana* and *H. camporum*—small size.

Initial selections for resistance to SALB in Brazil were done by Ford Company. During 1942–1945, the program was expanded and conducted in cooperation among Ford Company itself, Agronomic Institute of the North (IAN)—newly established at that time[1], and the Department of Agriculture of the United States [15].

[1] Currently CPATU—Center for Agricultural Research in the Humid Tropics.

The first step was to select matrices that had demonstrated resistance to the disease in Fordlândia. The selected resistant clones were known as Ford clones, designated by letter F (Ford selection), such as clone FA 1639, a clone of *H. brasiliensis* originated of ungrafted plants from seeds from Acre State, and clone F 4542, originated of seeds of *H. benthamiana* at upper Rio Negro [16].

During the administration of Ford Motor Company, breedings between Ford clones resistant to *M. ulei* and productive clones from the East received the acronym FX, e.g., FX 4037, originated from the selection of a seedling resultant of the breeding between F 4542 and PB 86. Breedings conducted in 1945 and subsequent years, under the auspices of the Agronomic Institute of the North, received the acronym IAN. The materials available for the breeding program consisted of oriental clones susceptible to *M. ulei*, such as PB 86, PB 186, Tjir 1, Tjir 16, AVROS 183, and AVROS 363, considered the best producer clones of the period, as well as primary clones of *H. brasiliensis*, selected from Fordlândia and Belterra, and clones of other species of rubber trees collected all over the Amazon Basin. Having resistant and productive material, an intraspecific breeding program was developed with the aim of associating, in one same plant, the characters desirable for dry rubber production and resistance to leaf blight. However, the lack of genetic diversity between parentals led to no manifestation of the hybrid vigor for the character of resistance to the pathogen [17].

In view of the great susceptibility of genotypes obtained by intraspecific breedings, other sources of resistant germplasm had to be searched in other species of the genus *Hevea*, aiming at interspecific breeding involving productive plants of *H. brasiliensis* and other plants resistant to the pathogen. Thus, plants representing the following species were collected and taken to Belterra: *H. spruceana*, *H. microphylla*, *H. guianensis*, and *H. pauciflora*. Hybrids from the breedings *H. brasiliensis* × *H. guianensis*, *H. brasiliensis* × *H. microphylla*, and *H. brasiliensis* × *H. spruceana* were discarded for not meeting the goals of latex production and resistance. Hybrids of *H. benthamiana* (especially clones F 4542) with *H. brasiliensis*, selected in Fordlândia, started to constitute basic material of resistance in successful genetic breeding programs [18].

Since then, thousands of plants were selected as resistant, of which only a small number had a good phenotypic value (what the plant exteriorizes) for the character dry rubber production. According to [19], hybrids of *H. pauciflora* × *H. brasiliensis* presented high resistance to the fungus *M. ulei* but low production of dry rubber—material recommended, in recent years, for the genetic-horticultural control of leaf blight by means of crown grafting [20]. While Brazilian researchers searched for resistant and productive materials, the Malaysian program by the Rubber Research Institute Malaysia (RRIM) focused only on obtaining *H. brasiliensis* clones of high productivity since the disease was not a concern because the pathogen was absent. They obtained a good performance with series 500, which was subsequently supplanted with clones of series 600. The latter is known worldwide, and RRIM 600 has become one of the most cultivated clones for its high productivity and plasticity. Currently, RRIM is in series 900. Brazil has received these clones as a payment for the genetic material supplied during the last years.

Assessing the behavior of rubber tree clones to leaf blight is extremely important for the establishment and success of rubber tree plantations. Silva [21] evaluated the behavior of 18 rubber tree clones against SALB. The evaluations were performed at 15-day intervals by removing 30 leaflets per tree. The disease was quantified based on the number of leaflets that were collected and classified according to the stages of the development of the disease and the type of injury. SALB occurred during the entire experimental period; however, the intensity of the disease varied in accordance with the resistance level of the clones and the time of year. The rubber clones FX 3864, RRIM 725, RRIM 711, IAC 300, and IAN 873 exhibited the highest resistance to leaf blight (**Figure 2**).

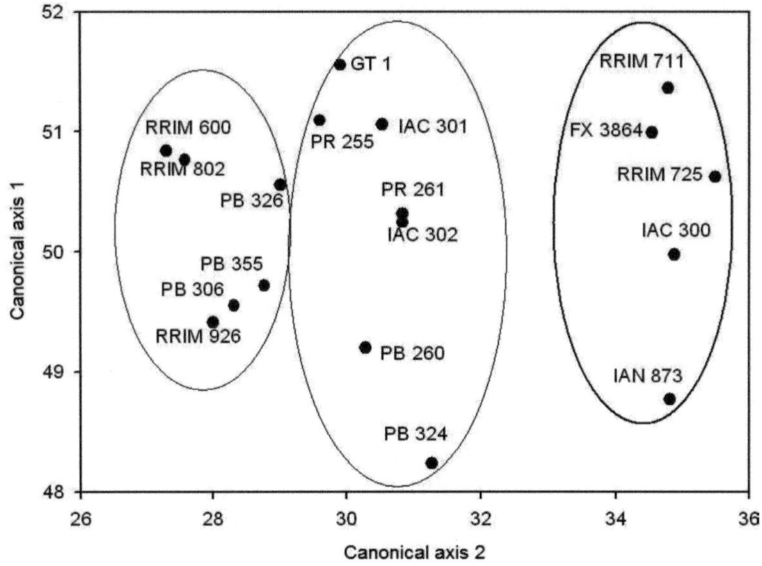

Figure 2.
Clustering of the studied rubber clones in relation to SALB (M. ulei) based on Mahalanobis distances [21].

5.3 Climate zoning

The Climatology Sector of Campinas Agronomic Institute developed a proposal of climate zoning for rubber by adopting the water balance method, that is, an accounting study involving the potential monthly values of rainfall and evapotranspiration of each region and their correlation with the intensity of leaf blight symptoms. The zoning divided the Amazon region into four distinct ecological zones, while the remaining Brazilian regions were distributed into other seven zones (**Table 2**).

5.4 Management of SALB per Brazilian climate region

5.4.1 Dry regions and spatial evasion

One of the major measures related to the management of leaf blight in Brazil includes cultivation at sites unfavorable to the development of the pathogen, using the general principle of evasion control (geographic evasion or evasion in space), popularly known as "scape areas."

According to the climate zoning (**Table 2**), a vast area in Brazil is considered preferential, with potential for rubber cultivation (region A, A1, and AM4) and well-defined dry season coinciding with the leaf shedding period of plants (clones of uniform "deciduous" habit, from *H. brasiliensis*), without risks of epidemics. This area corresponds to 2/3 São Paulo State, ¼ Mato Grosso State, ¼ Mato Grosso do Sul State, Goiás State and Minas Gerais State, besides the South of Pará, and part of Tocantins and Maranhão States. Thus, new borders were opened for rubber cultivation in the country, showing the way to self-sufficient natural rubber production. An example is the state of São Paulo, which presently counts on ~50,000 ha mostly covered with eastern clones of high productivity (e.g., RRIM 600, PB 235, and PR 255), without any concern about leaf blight, and is responsible for more than 50% of rubber produced in the country. Other examples are the south of Mato Grosso (A1), which has the largest continuous rubber tree plantation in the country, 8500 ha, and the south of Maranhão (AM4), where plantations developed very well, without leaf blight.

5.4.2 Humid regions and temporal evasion (avoidance)

In regions B and B1, with phytosanitary restrictions for cultivation, preference should be given to national clones of *H. brasiliensis* with resistance and uniform leaf shedding, which usually occurs during the period with scarce rainfall and temperature below optimal for infection by *M. ulei*. For these regions, there is great performance of clones with early leaf shedding, resulting in long hibernation periods, e.g., IAN 873, FX 4098, FX 2261, FX 985, and FX 3864. Clone MDF 180 has marked phenol production by infected leaflets. This inhibits the sexual phase of *M. ulei*, naturally breaking the pathogen's cycle, which, associated with the clone's deciduous habit, significantly reduces epidemics [22]. Using such materials favors once again the application of the evasion principle, in this case, evasion in time or avoidance, a term first proposed by [23], since local conditions favor the disease and clones should be selected for their phenological qualities [24], besides productivity.

Hybrid clones, from the breeding of *H. brasiliensis* with *H. benthamiana*, erroneously recommended in the 1970s in supporting programs (PROBOR I, II, and III), should be avoided due to their irregular habit of leaf shedding, a negative aspect for disease control, since it does not allow a break in the pathogen's life cycle and does not reconstitute vertical resistance to materials. Eastern cultivars, in general originated from *H. brasiliensis*, were selected for latex production; therefore, they are highly susceptible and consequently not recommended for these regions.

5.4.3 Super humid regions and crown grafting

In the super humid and humid Amazon region (AM1, AM2, and part of AM3), the use of resistant cultivars showing uniform leaf shedding is not sufficient to control the disease, since both leaf wetness period and temperature are high all over the year, favoring infection. Thus, plantations in continuous areas should involve crown grafting with hybrid cultivars of *H. pauciflora*, a species that has remained highly resistant to leaf blight during all these years. In this case, the adopted seedlings should be of the tricompound type or with double graft; they should also be constituted of vigorous and rustic rootstock, a first graft with productive clone, which will result in the future panel, and a third crown graft with these resistant hybrids at a height of 2.5 m, which will constitute the future crown of the tree [25].

5.4.4 Super humid regions and neoextractivism

In view of the low inoculum pressure existing in native rubber trees in the forest (see text above), besides crown grafting, the philosophy of neoextractivism can be adopted to establish plantations in these regions, i.e., extractivism with sustainability and technology, a management proposal tested in Acre State at Chico Mendes Extractive Reserve (RESEX) and Caquetá Extractive Settlement with plantations in small areas, where subsistence farming is practiced (from 1 to 1.5 ha), i.e., forest enrichment with greater density of productive rubber trees in small plantations from seeds (ungrafted) or polyclonals, associated with other species of interest such as "açai," cocoa, "cupuaçú," banana, coffee, etc., or even leaving the plants in the regenerating forest.

To constitute polyclonals, several clones of *H. brasiliensis* can be used; they should present uniform leaf shedding in the adult phase and be previously selected for resistance to different *M. ulei* races, in a larger spacing, to make up from 250 to 300 plants per hectare, surrounded by forest, which are named "highly productivity islands" (HPLs) due to the high latex production of these clones and the possibility of association with other species that allow economic use, improving the income and the rubber tappers' life condition, without leaving their activity [26]. To keep

Etiological agent	Disease	No. Product s	Commercial Product	Active Ingredient (Chemical Group)
Microcyclus ulei	South American Leaf blight of rubber	9	Bravonil 750 WP	Chlorothalonil (Isophthalonitrile)
			Cercobin 700 WP	Thiophanate-methyl (Benzimidazole)
			Cobre Atar BR	Cuprous Oxide (Inorganic)
			Daconil BR	Chlorothalonil (Isophthalonitrile)
			Dacostar 750	Chlorothalonil (Isophthalonitrile)
			Metiltiofan	Thiophanate-methyl (Benzimidazole)
			Redshield 750	Cuprous Oxide (Inorganic)
			Topsin 500 SC.	Thiophanate-methyl (Benzimidazole)
			Topsin 700	Thiophanate-methyl (Benzimidazole)

Table 3.
Fungicides registered for the control of South American leaf blight in Brazil [29].

such "islands" protected from SALB epidemics, their number should not be >8, at every 400 ha, which compose a setting of native rubber trees, for RESEX, or up to two islands for every 100 ha, corresponding to one lot of the Extractive Settlement, always keeping the native forest intact in the surroundings since the native forest, in this case, acts as a natural barrier against fungal dispersion.

5.5 Additional practice to the management of South American leaf blight

5.5.1 Chemical control

As to chemical control in Brazil, there are efficient fungicides to reduce the disease intensity. Examples are the active principles thiophanate-methyl and triadimefon, which act on fungal stromata, making them sterile [27], as well as chlorothalonil, which has high residual potential while triadimefon has a healing effect. Furtado [28] found that weekly application of mancozeb and biweekly application of fenbuconazole and myclobutanil were efficient in controlling the disease. The fungicides registered for the disease control are listed below (**Table 3**).

5.5.2 Biological control

Biological control of the pathogen by the hyperparasite fungus *Dicyma pulvinata* represents a potential control measure. Studies conducted in greenhouse, nursery, clonal garden, and still young definitive plantation (4–5 years old) have shown efficient pathogen control under Amazon conditions. This hyperparasite reduces the primary inoculum since it prevents the stromatic phase of *M. ulei*, making unviable the production and the dissemination of ascospores [25]. This fungus was found parasitizing lesions in a clonal garden at the north coast of São Paulo State.

6. Potential impact of global climate changes on the spatial distribution of South American leaf blight in Brazil

The climate changes of the last decades have attracted the attention of the different segments of society, especially concerning their causes and consequences [30]. Of all economic sectors, agriculture has the greatest dependency on environmental conditions, especially climate. The impacts on plant diseases are differently expressed, and emphasis should be given to the effects of damages caused by the diseases on geographic distribution, efficiency of control methods, and remaining organisms that interact with the plant [31–36].

Among the phytosanitary problems of the country, leaf blight, caused by *M. ulei*, constitutes the major factor of productivity loss for rubber tree plantations in Brazil. Temperature and relative humidity have great influence on the rubber culture and its productivity, as well as on the geographic and temporal distribution of diseases [30]. Global climate changes constitute a serious threat to the Brazilian phytosanitary scenario since they can promote significant changes in the occurrence and severity of diseases affecting agricultural and forest plants. Based on the importance of this disease for the economic scenario of the country, the present study aimed to evaluate possible impacts of global climate changes on the spatial distribution of leaf blight in Brazil.

To elaborate spatial distribution maps of the disease, [37] regarded the monthly data for the period 1961–1990 as the current data of mean temperature and relative humidity, as indicated by the Intergovernmental Panel on Climate Changes (IPCC) and by the World Meteorological Organization, to characterize the future climate for every month, forecasts obtained from IPCC were used with the model developed by Hadley Center for Climate Prediction and Research (HadCm3). The scenarios used to obtain future projections were A2 and B2, centered in the decades of 2020 (period between 2010 and 2039), 2050 (period between 2040 and 2069), and 2080 (period between 2070 and 2099) [38]. Scenario A2 has high rates of greenhouse gas emission, i.e., it assumes the maintenance of the current emission patterns. Scenario B2 has less emission and shows more optimistic characteristics relative to scenario A2 [39].

Adopting the model HadCM3, data were interpolated by the method of inverse square of distance. Then, a mask delimitating the continents was applied on the maps and subsequently the area corresponding to Brazil was cut from the georeferenced data. For each month, maps containing data of temperature and relative humidity were generated considering the current climate situation and forecasts for the decades of 2020, 2050, and 2080, for scenarios A2 and B2.

Using the overlay techniques, maps containing the spatial distribution of areas where each climate element favors the development of the pathogen were elaborated. From the maps of mean temperature and monthly relative humidity for scenarios A2 and B2, in the current and future period, monthly distribution maps of areas favorable or not to the disease were elaborated, using classes defined based on the available epidemiological data about the effects of temperature and relative humidity on the development of the disease (**Table 4**) [8, 40–43].

The maps of risk areas for leaf blight elaborated for the future scenarios indicate that, in general, there will be a reduction in the area highly favorable and favorable to the disease in the country, relative to the current climate, for both scenario A2 and scenario B2 in **Figure 3**. Such a reduction is projected for either the period of greatest favorability to the disease (January, February, and March) or the least favorable period (July to October).

The major change in climate that is responsible for this result is probably the reduction in the mean relative humidity to levels unfavorable to the disease, i.e., values below 65%. In general, the reductions in the disease incidence were more pronounced for scenario A2 than for scenario B2. Scenario A2 predicts greater reductions in humidity than scenario B2, resulting in conditions less favorable to SALB.

Considering the current and the future scenarios, a reduction in the percentage of highly favorable and favorable areas is expected, as well as an increase in the areas relatively favorable and unfavorable to occurrence of SALB. Some regions of

Favorability Classes	Temperature (°C)	Relative Humidity (%)	Description
1	24 to 28	>90	Highly Favorable
2	20 to 24	>80	Favorable
2	24 to 28	80 to 90	Favorable
3	16 to 20	>80	Relatively Favorable
3	16 to 28	65 to 80	Relatively Favorable
4	<16 and >28	<65	Unfavorable

Table 4.
Classes of favorability to the development of SALB according to temperature and relative humidity intervals.

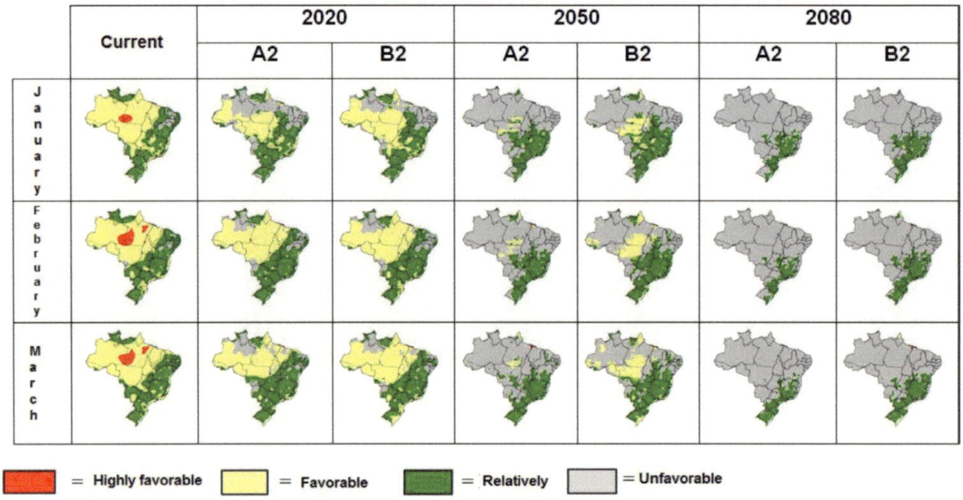

Figure 3.
Maps of climate favorability to South American leaf blight for January, February, and March in the current (1961–1990) and future scenario (years 2020, 2050, and 2080), considering scenarios A2 and B2 [36].

the country will become more apt to cultivation than others, which may trigger the emergence and/or better development of some new plantation regions. The months with higher temperatures will be more favorable to SALB; on the other hand, colder months are considered unfavorable to the development of this disease under the current climate conditions, remaining constant for future projections. The obtained knowledge associated with the development of disease prediction models can constitute important tools in the integrated management of SALB.

Author details

Edson Luiz Furtado[1], Willian Bucker Moraes[2], Waldir Cintra de Jesus Junior[3]*, Breno Benvindo dos Anjos[4] and Lilianne Gomes da Silva[4]

1 Forest Pathology, Department of Plant Protection, School of Agronomical Sciences, São Paulo State University—UNESP, Botucatu, SP, Brazil

2 Department of Agronomy, Federal University of Espírito Santo—UFES, Alegre, ES, Brazil

3 Nature Science Center, Federal University of São Carlos—UFSCar, Buri, SP, Brazil

4 Federal University of Espírito Santo—UFES, Alegre, ES, Brazil

*Address all correspondence to: wcintra@ufscar.br

References

[1] Gasparotto L. Epidemiologia do mal das folhas (*Microcyclus ulei* P. Henn. V. arx) da seringueira (*Hevea* spp) [thesis (doctorate)]. Viçosa: Universidade Federal de Viçosa; 1988. 124p

[2] Furtado EL, Trindade DR. Doenças da seringueira. In: Kimati H et al., editors. Manual de Fitopatologia: Doenças das culturas, V.II, 4. Ed. Piracicaba: Agronômica Ceres; 2005. pp. 559-569. ISBN: 85-318-0043-9

[3] Gasparotto L, JCR P, Ferreira FA, Furtado EL, Santos AF. Doenças da seringueira no Brasil. Brasilia-DF: EMBRAPA; 2012. p. 255. ISBN: 978-85-7035-097-8

[4] Hora Júnior BT, Macedo DM, Barreto RW, Evans HC, Mattos CRR, Maffia LA, et al. Erasing the past: A new identity for the Damoclean pathogen causing South American leaf blight of rubber. PLoS ONE. 2014;**9**(8):e104750. DOI: 10.1371/journal.pone.0104750

[5] Chee KH. Factors affecting discharge, germination and viability of spores of *Microcyclus ulei*. Transactions of the British Mycological Society. 1976;**66**:499-504. DOI: 10.1016/S0007-1536(76)80221-5

[6] Camargo AP, Cardoso RMG, Schmidt NC. Comportamento ecológico do mal das folhas da seringueira nas condições climáticas do Planalto Paulista. Bragantia. 1967;**26**:1-18

[7] Furtado EL, Kageyama PY, Souza AD, Costa JD. Ilhas de alta produtividade (Iap): Uma proposta para produtividade com biodiversidade nas reservas extrativistas do Acre. In: Seminário de Ciência e Desenvolvimento Sustentável. São Paulo, SP: Anais; 1997. pp. 29-31

[8] Chee KH. Assessing suceptibility of *Hevea* clones to *Microcyclus ulei*. Annals of Applied Biology. 1976;**84**:135-145. DOI: 10.1111/j.1744-7348.1976.tb01743.x

[9] Junqueira NTV. Variabilidade fisiológica de *Microcyclus ulei* (P. Henn.) v. Arx. [thesis (doctorate)]. Viçosa: Universidade Federal de Viçosa; 1985. 135p

[10] Ortolani AA. Agroclimatologia e o cultivo da seringueira. In: Simpósio sobre a cultura da seringueira no estado de São Paulo. Piracicaba, SP: Anais; 1986. pp. 11-32

[11] Honorato Júnior JJ. South American leaf blight of rubber tree: Dynamics of pathogen inoculum, progress and losses, in three topographical conditions [thesis (masters)]. Viçosa: Universidade Federal de Viçosa; 2010. 93p

[12] LMVD C, Jones C. Zona de Convergência do Atlântico Sul. In: Cavalcanti IFDA et al., editors. Tempo e Clima no Brasil. São Paulo: Oficina de Textos; 2009. p. 463

[13] Furtado EL, Cunha AR, Alvares CA, Bevenuto JAZ, Passos JR. Ocorrência de epidemia do mal das folhas em regiões de "escape" do Brasil. Arquivos do Instituto Biológico. 2015;**82**:1-6. DOI: 10.1590/1808-657000882013

[14] Ortolani AA, Pedro Junior MJ, Alfonsii RR, Camargo MBP, Brunini O. Aptidão agroclimática para a regionalização da heveicultura no Brasil. In: Seminário de Recomendação de Clones de Seringueira, Brasília, 1982. MIC/SUDHEVEA: Anais; 1983. pp. 19-28

[15] Towsend Junior CHT. Progress in developing superior *Hevea* clones in Brazil. Economic Botany. 1960;**14**:198-196

[16] Rands RD, Polhamus LG. Progress report on the cooperative rubber

development programme in Native America. Circular United States. Department of Agriculture. 1955;**976**:30

[17] Baptiste EDC. Recent progress in Malaya in the breeding annual selection clones of *Hevea brasiliensis*. In: International Horticulture Congress. London: London Proceedings; 1952. pp. 1100-11221

[18] Valois ACC. Melhoramento genético da seringueira. In: Curso de especialização em heveicultura. Belem: EMBRAPA/CNPSD/FAP; 1978. p. 24

[19] Pinheiro E, Libonati VF. O emprego da *Hevea pauciflora* M.A. como fonte genética da resistência ao mal-das-folhas. vol. 1(1). Rio de Janeiro: Polímeros; 1971. pp. 31-40

[20] Moraes VHF. Fisiologia da seringueira. In: Curso de especialização em heveicultura. Belém: SUDHEVEA/FCAP; 1985. p. 40

[21] Silva LG, Jesus Junior WC, Souza AF, Alves FR, Furtado EL. Performance of different rubber tree clones against South American leaf blight (*Microcyclus ulei*). Forest Pathology. 2014;**44**:211-218. DOI: 10.1111/efp.12084

[22] Sambugaro R, Furtado EL, Rodella RA, Mattos CRR. Anatomia foliar de seringueira (*Hevea* spp.) e desenvolvimento da infecção por *Microcyclus ulei*. Summa Phytopatothologica. 2004;**30**:44-50

[23] Evitação MJOM. Forma de defesa das plantas contra patógenos que deve ser melhor compreendida e explorada. Summa Phytopathologica. 1990;**16**:77-83

[24] Furtado ELacr [thesis (doctorate)]. Piracicaba: Escola Superior Agronomia Luiz de Queiroz, Universidade de São Paulo; 1996

[25] Gasparotto L, Trindade DR, Silva HM. Doenças da seringueira. Manaus: Embrapa-CNPSD; 1984. p. 71

[26] Kageyama PY. Extractive reserves in Brazilian Amazônia and genetic resources conservation. Tenth world Forestry Congress in Paris; September 1991. pp. 115-119

[27] Brignani Neto F, Furtado EL, Cardoso RMG, Rolim PRR, Oliveira DA. Efeito de fungicidas sistêmicos no ciclo biológico de *Microcyclus ulei*, agente da queima da folha da seringueira (*Hevea* spp.). Summa Phytopatothologica. 1991;**17**:238-246

[28] Furtado EL, Menten JOM, Carvalho JC, Godoy Junior G. Ação de fungicidas inibidores de demetilação, na síntese de ergosterol, no controle do mal das folhas da seringueira. Fitopatologia Brasileira. 1995;**20**:203-208

[29] AGROFIT. Brasília, DF: Ministério da Agricultura, Pecuária e Abastecimento; 2017. Disponível em: http://agrofit.agricultura.gov.br/agrofit_cons/!ap_praga_detalhe_cons?p_id_cultura_praga=4712 [Disponível em: 06 September 2018]

[30] Ghini R, Hamada E. Mudanças climáticas: Impactos sobre doenças de plantas no Brasil. Jaguariúna: Embrapa Meio Ambiente; 2008. p. 332

[31] Atkinson D. Global climate change: Its implication for crop protection. British Crop Protection Council Monograph No. 56; Surrey: BCPC; 1993

[32] Chakraborty S, Tiedemann AV, Teng PS. Climate change: Potential impact on plant diseases. Environmental Pollution. 2000;**108**:317-326

[33] Chakraborty S. Effects of climate change. In: Waller JML, Waller SJ, editors. Plant Pathologist's Pocketbook. Wallingford: CAB International; 2001. pp. 203-207

[34] Chakraborty S. Potential impact of climate change on plant-pathogen interactions. Australasian Plant Pathology. 2005;**34**:443-448

[35] Jesus Junior WC, Cecilio RA, Valadares Junior R, Cosmi FC, Moraes WB, Alves FR, et al. Aquecimento global e o potencial impacto na cultura e doenças do mamoeiro. In: Costa NA, Costa AFS, editors. Congresso Brasileiro de Fruticultura Tropical. Vitória: INCAPER; 2007. pp. 1-36

[36] Jesus Junior WC, Valadares Júnior R, Cecílio RA, Moraes WB, Vale FXR, Alves FR, et al. Worldwide geographical distribution of Black Sigatoka for banana: Predictions based on climate change models. Scientia Agricola. 2008;**65**:40-53. DOI: 10.1590/S0103-90162008000700008

[37] Silva LG. Comportamento de clones de seringueira ao mal das folhas e potencial impacto das mudanças climáticas globais na ocorrência da doença [thesis (másters)]. Alegre: Universidade Federal do Espírito Santo; 2010

[38] IPCC. Climate Change 2007: The Physical Science Basis: Summary for Policymakers. Geneva: IPCC. 2007. 18p. Disponível em: http://www.ipcc.ch/SPM2feb07.pdf [Acesso em: 10 June 2009]

[39] IPCC. Climate Change 2001: The Science Basis IPCC WGI, TAR. New York: Cambridge University Press; 2001. p. 881

[40] Langford MH. South American Leaf Blight of *Hevea* Rubber Trees. Washington, D.C: USDA; 1945. p. 31

[41] Holliday P. South American Leaf Blight (*Microcyclus ulei*) of *Hevea brasiliensis*. Farnham Royal: CAB; 1970. p. 31

[42] Gasparotto L, Zambolim L, Maffia L, Vale FXR, Junqueira NTV. Efeito da temperatura e da umidade sobre a infecção de seringueira por *Microcyclus ulei*. Fitopatologia Brasileira. 1989;**14**:38-41

[43] Gasparotto L, Zambolim L, Ventura JA, Costa H, Vale FXR, Maffia LA. Epidemiologia do mal das folhas da seringueira no Estado do Espírito Santo. Fitopatologia Brasileira. 1991;**16**:180-184

Chapter 7

Fusarium Wilt in Banana: Epidemics and Management Strategies

Fatin Nadiah Jamil, Chu-Nie Tang, Noor Baity Saidi, Kok-Song Lai and Nadiya Akmal Baharum

Abstract

Fusarium wilt, caused by *Fusarium oxysporum* f. sp. *cubense* (Foc), is one of the most threatening fungal diseases affecting banana plantations across the globe. It was first discovered in Australia in 1874 and has now spread to numerous different regions in the world hinting at the persistency of the pathogen. Various management strategies have been devised aiming mainly on improving the plant's tolerance or suppressing the infection. Fungicide is commonly used to control the disease spread, but it does not provide total protection to the plants besides displaying selective effectiveness on certain Foc strains. Alternatively, farmers apply crop rotation, rice hull burning, biological soil disinfestation, and compound-supplemented soil in their banana plantations. Studies have also shown that certain biocontrol agents manage to curb the disease threat. Selection of somaclonal variants and genetic manipulation via induced mutagenesis and transformation are also among the alternatives that have been implemented in producing *Fusarium*-tolerant and *Fusarium*-resistant banana plants. This chapter will describe *Fusarium* epidemics in banana, the effectiveness and challenges of different management approaches, as well as the future alternatives that can be adopted by taking advantages of the latest advances in omics technologies.

Keywords: disease control, epidemiology, *Fusarium oxysporum* f. sp. *cubense*, Foc, fungal disease, soilborne

1. Introduction

Fusarium wilt is the most destructive fungal disease affecting banana plantation across the globe [1–4]. It is caused by a type of soilborne fungal called *Fusarium oxysporum* f. sp. *cubense* (Foc). The pathogen penetrates through banana roots and dominates the vascular tissues, which disrupts the dissemination of necessary nutrients from roots to the upper parts of the plants [5–8]. It was first discovered in Australia [9] and was later followed by reports from Tropical America (Costa Rica and Panama) in 1890 [10]. Despite the first occurrence in Australia, the disease was believed to have originated in Southeast Asia [10–13]. Drastic increase in the number of new disease records occurred in the early 1900s causing over USD 2.3 billion loss in the last century [14], most of which described the damages in the export plantations of "Gros Michel" variety affected by race 1 of Foc [1, 10, 15].

In the recent years, Foc Tropical race 4 (FocTR4) emerges as the most virulent strain causing epidemics in Taiwan, Peninsular Malaysia, Indonesia, the Philippines [16], China, Jordan, Mozambique, Lebanon, Pakistan [6, 17–20], Australia [21, 22], Vietnam [23], Laos [24], Myanmar [8], and Israel [25]. The latest outbreak of FocTR4 has been confirmed in the Americas affecting the most popular commercial variety which could have jeopardized banana production for decades [26]. According to several unofficial reports, about 15,500 hectare (ha) of plantation areas in the Philippines, 40,000 ha in China, and 80% of production area in the Jordan Valley had been jeopardized by FocTR4 [14]. Several management strategies have been executed in controlling the spread of *Fusarium* wilt. However, they did not manage to provide long-term solutions to the problem. Current cutting-edge technologies particularly through omics studies may provide better alternatives to solve this long-running issue. Here, we discuss the current knowledge on *Fusarium* wilt with special emphasis on current disease management approaches and rising new omics technologies that can be adopted.

2. Symptoms of *Fusarium* wilt

Fusarium oxysporum f. sp. *cubense* (Foc) causes a typical wilt syndrome on the infected banana plants accompanied by the necrosis and rotting of roots, rhizome, and pseudostem vessels. The most typical symptoms become visible in susceptible banana plants after the appearance of initial external symptoms such as pale green streaks on the base of the petiole and the brown-reddish discoloration of the vessels under the epidermis of the petiole (**Figure 1**). These symptoms occur between 2 and

Figure 1.
External symptoms of Fusarium wilt in cv. "Berangan" banana cultivated in a plantation located in Selangor state, Malaysia (3°48'10.1"N 100°50'42.5"E). (a) Split pseudostem. (b) Skirting of wilted leaves. (c) Leaves with yellow and brown streak. Red arrows indicate external symptoms observed.

5 months after infection of roots [10]. Foc and other members of *Fusarium oxysporum* (Fo) species complex produce a type of phytotoxin called fusaric acid (FA), which has been suggested as the cause of leaf chlorosis symptom [27].

Discoloration observed in the cross and longitudinal section of the rhizome, pseudostem, as well as petiole is among the internal symptoms of *Fusarium* wilt. Golden discoloration may also be observed in infected rhizome. In more severe infection, the rhizome displays total discoloration accompanied by sticky texture and rotten smell (**Figure 2**). Meanwhile, healthy banana plants produce fresh green leaves, clear cross section of petiole with no brown ring as well as white and healthy rhizome (**Figure 3**).

The first internal symptom of the disease occurs in the hair roots which are the initial sites of infection. The infection later progresses to the rhizome. In a more prominent attack, the pathogen may colonize the cortex as well as the vascular bundle of the pseudostem. The pathogen passes through the affected vessels to the new growing shoot [28]. Appearance of brown-reddish discoloration in the internal vessels of the pseudostem confirms that the pathogen has already invaded the pseudostem. The oldest leaf sheaths may also show brownish streaks [1]. In general, no internal symptom is observed in the fruits.

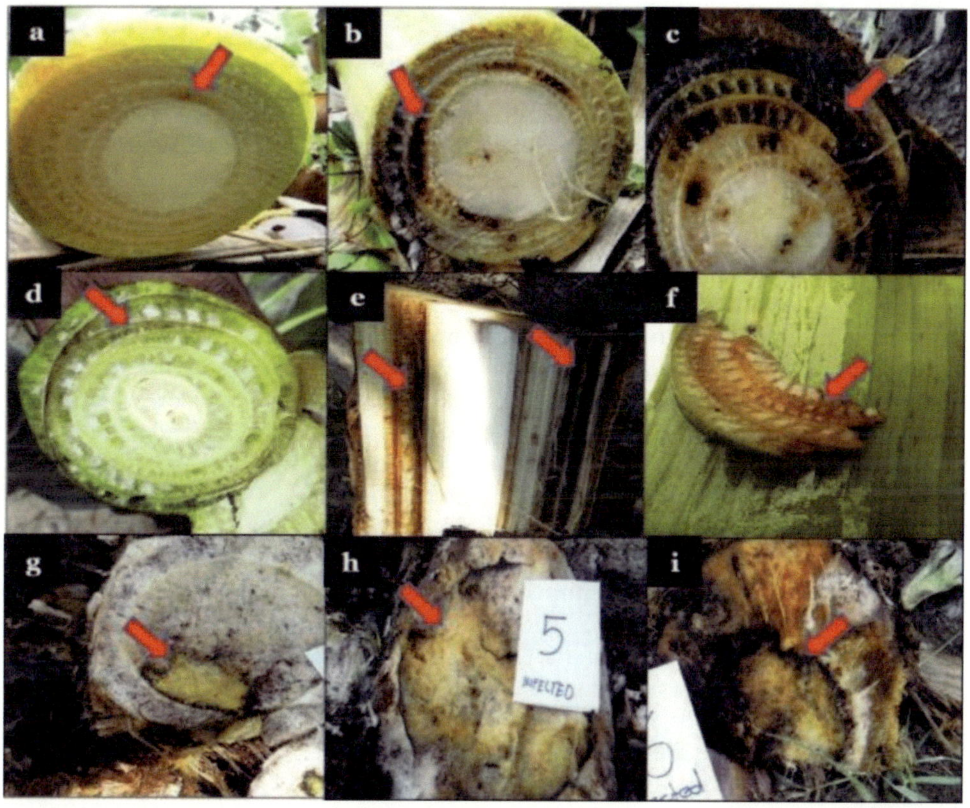

Figure 2.
Internal symptoms of Fusarium wilt in cv. "Berangan" banana cultivated in a plantation located in Selangor state, Malaysia (3°48′10.1″N 100°50′42.5″E). (a) Cross section of pseudostem showing minor discoloration. (b) Pseudostem showing moderate vascular discoloration. (c) Severe internal discoloration of the infected pseudostem. (d) Cross section of infected petiole showing minor discoloration. (e) Longitudinal section of pseudostem showing infected vascular bundle. (f) Petiole showing severe discoloration. (g) Infected rhizome showing minor golden discoloration. (h) Rhizome showing moderate discoloration. (i) Total discoloration in severely infected rhizome producing sticky texture and rotten smell. Red arrows indicate discoloration sites.

Figure 3.
Healthy banana without Fusarium wilt symptoms. (a) Fresh green leaf. (b) Clear cross section of petiole with no brown ring. (c) White and healthy rhizome. (d) Cross section of pseudostem with no brown ring.

3. *Fusarium* wilt infection

Over the years, researches have been carried out to investigate the infection process and disease cycle of *Fusarium* wilt. It was found that mycelia and conidia are produced after 6–8 hours of chlamydospore germination, while new chlamydospores will be produced after 2–3 days [28–30]. The infection is not initiated through the main root but instead takes place through secondary or tertiary feeder roots [31]. According to Beckman [32, 33], the pathogen infects the roots of both susceptible and resistant cultivars, but the infection of vascularized fragment of the rhizome is more prominent in the susceptible genotypes. In responding to the infection, tyloses, gums, and gel are produced in xylem lumen. Most of the infections are blocked, but some of them become systemic and pass through rhizome and pseudostem. The pathogen may mobilize through many different vascular pathways starting from the roots [30, 34, 35]. However, pathogen colonization in the infected rhizome is deemed the most efficient [28]. As the disease advances, the pathogen moves out of the vascular system to the adjacent parenchyma forming conidia and chlamydospores which are released to the soil when the plant died. These spores may remain dormant in the soil for years [36–38]. Ploetz [17] suggested that the pathogen's persistent survivability in the soil is contributed by the presence of living banana host.

4. Classification

Foc is a genetically and pathogenically diverse fungus. Over 20 vegetative compatibility groups (VCGs) and diverse evolutionary lineages are recognized in the global populations of the pathogen [1, 39–45]. Foc can also be classified according to races depending on the group of cultivars they affected. To date, four races of Foc have been recognized [46–48].

Race 1 of Foc which was responsible for the "Gros Michel" epidemics also attack "Maqueno" (Maia Maoli-Popoulu subgroup, AAB), "Silk," "Pome" AAB, and "Pisang Awak" ABB. Race 2 of Foc is known to affect cooking bananas and plantains such as "Bluggoe" (ABB). On the other hand, race 3 of Foc is described as a pathogen of *Heliconia* spp., which is a tropical American banana relative [49]. It was also found to have minor impact on "Gros Michel" and seedlings of *M. balbisiana*. Race 3 Foc has not been reported since Waite's [49] work, and no voucher specimens of the pathogen exist. Of these, race 4 Foc is considered the most virulent group which does not only affect economically important cultivars like "Cavendish" but also race 1- and race 2-susceptible varieties [48, 50–52].

Race 4 Foc can be further divided into Tropical (TR4) and Subtropical (SR4) variants [50, 53]. Foc Subtropical race 4 (FocSR4) isolates cause infection on plants that are grown in an unconducive environment such as cool temperature, poor soil, and under stress condition. It also has an ability to infect banana plants regardless of the predisposing conditions [54]. Foc Tropical race 4 (FocTR4) cases have been reported in Taiwan, Indonesia, Malaysia, China, the Philippines, and Northern Australia [6]. The early outbreaks of FocTR4 in China and the Philippines were not taken seriously that they slowly developed into destructive and uncontrollable problems [38]. Recent outbreaks of FocTR4 in Mozambique and Jordan were claimed as the first reported FocTR4 cases outside the common Asian and Australian regions [6] (**Table 1**).

5. Geographical distribution

Year	Country	Description
1874	Australia	Dr. Joseph Bancroft was the first person to discover that *Fusarium* wilt was caused by a type of fungus. This was revealed using microscopic examination
1890	Panama and Costa Rica	First report in Panama and Costa Rica
1940	N/A	Snyder and Hansen suggested the pathogen be renamed as *Fusarium oxysporum* f. sp. *cubense*
1890–mid 1950s	Central and Southern America	*Fusarium* epidemic (race 1) annihilated more than 50,000 hectares of cv. "Gros Michel" plantations. "Gros Michel" cultivar was replaced with "Cavendish" as the major cultivar for trade
1911	India	*Fusarium* wilt was first reported in different countries
1916	Java, Indonesia	
1920	Philippines	
1953	Malaysia	
1967	Taiwan	
1990–present	N/A	Foc Tropical race 4 (FocTR4) severely affected the banana productions worldwide
1994	N/A	First FocTR4 case on cv. "Cavendish"
Mid-1990s	Malaysia and Indonesia	Severe *Fusarium* epidemics in Malaysia and Indonesia
1996	Panyu, Guangdong, China	First report on infected Cavendish
1997–1999	Near Darwin (northern territories), Australia	Three FocTR4 outbreaks in sites

Year	Country	Description
2000	Southern China	Epidemic severely affected cv. "Cavendish" plantations
2002	China	FocTR4 annihilated about 60,000 ha of banana plantation
2013	Oman, Jordan, and Mozambique	FocTR4 had spread outside the common Asia regions
2015	Pakistan and Lebanon	First report of FocTR4 causing Panama disease in Cavendish bananas in Pakistan and Lebanon
2018	Vietnam	First report of *Fusarium* wilt on Cavendish bananas caused by FocTR4 (VCG 01213/16) in Vietnam
2018	Laos	First report of FocTR4 (VCG 01213/16) associated with Cavendish bananas in Laos
2018	Myanmar	FocTR4 incidence was reported in Myanmar
2018	Israel	First report of FocTR4 causing *Fusarium* wilt of Cavendish bananas in Israel
2019	America	Foc TR4 incidence was reported in the Americas

Table 1.
Significant events that took place after the discovery of Fusarium wilt [8, 10, 14, 17, 18, 23–26, 55, 56].

6. Management control

The effectiveness of *Fusarium* disease management strategies has been hampered by several limitations including extremely poor fertility of certain economically important banana varieties (e.g., "Cavendish"). This has particularly hindered the improvement efforts through conventional breeding [57]. Several approaches that have been practiced to curb the infection of *Fusarium* wilt in banana are biological control, chemical control, cultural control, physical control, quarantine, exclusion and personnel awareness, breeding programs, selection of somaclonal variants, and genetic modification via transgenic approach and mutagenesis.

6.1 Biological control

Due to the continuous spread of *Fusarium* wilt, biological control offers a commendatory disease management approach [58–61]. Diverse microbes such as *Pseudomonas fluorescens*, *Trichoderma viride*, *Bacillus* spp., *Serratia* spp., *Pseudomonas aeruginosa*, *Streptomyces violaceusniger*, *Y-Proteobacteria*, *Bacillus subtilis*, *Pseudomonas* spp., and nonpathogenic *Fusarium* had been tested against *Fusarium* wilt disease in which most of the published works resulted from in vitro assays or short-term greenhouse studies [34, 59, 62–72]. Some Indian medicinal plants such as *Calotropis gigantea* L., *Centella asiatica* L., *Ocimum sanctum* L., *Piper betle* L., and *Vitex negundo* L. have also been tested in vitro as biological control against Foc. Among those, *P. betle* L. extract exhibited the highest antifungal activity against the tested plant pathogen *Foc*, followed by *V. negundo*, *C. gigantea*, *C. asiatica*, and *O. sanctum* extracts [70]. These results proved that several plant extracts exhibit antifungal activity and have potency to reduce mycelium growth under both greenhouse and field conditions [62]. Other studies reported on an application of biosynthesized silver nanoparticles to control pathogenic fungi [73]. It has also been shown that a community of *Gammaproteobacteria* is an indicator species of healthy banana plants on *Fusarium* wilt-infested and healthy fields in Nicaragua and Costa Rica [74]. These researches provide new insights on using organisms as bioindicators. Nevertheless, there were very few studies done on the long-term biocontrol efficacy of *Fusarium* wilt in banana [2, 75]. For example, Thangavelu and Jayanthi [76] reported the field application of a selected nonpathogenic (np) Fo isolate, Ro-3, which reduced the severity of the disease up to 89% after three rounds of treatments.

On the other hand, Belgrove et al. [77] reported that neither the nonpathogenic *F. oxysporum* and *P. fluorescens* nor combinations thereof reduced the development of *Fusarium* wilt significantly.

6.2 Chemical control

Nel et al. [78] reported that prochloraz and propiconazole notably inhibited mycelial growth. In the same study, it was found that the disease symptoms of *Fusarium* wilt can be controlled up to 80.6% when benomyl and the demethylation-inhibiting fungicide were applied via root dip treatment. Besides, certain quaternary ammonium compounds were also reported as potent sterilants against Foc [78]. Similar to biological control, future field evaluations on the efficacy of the fungicides against *Fusarium* wilt are deemed necessary.

Test injections of fungicides into *Fusarium* banana plants had also been performed. In a study conducted by Lakshmanan et al. [79], rhizome of cv. "Rasthali" was injected with 2% carbendazim (Bavistin 50 WP) using hypodermic syringe. The upper portion of the rhizome was exposed by removing a portion of soil, and a slope hole at angle 45° was made before it was injected with the fungicide. The method managed to reduce the disease incidence up to 13.5% at the time of harvest.

In another study, Herbert and Marx [80] reported that the injection with carbendazim and several other fungicides did not manage to prevent the *Fusarium* wilt infection of banana in South Africa. In Australia, pseudostems of cv. "Williams" that were injected with 20% potassium phosphonate showed some effectiveness against Foc. However, the results were inconsistent and uncertain (Pegg, unpublished).

Herbert and Marx [80] also reported another treatment against *Fusarium* wilt using soil fumigation treated with methyl bromide. However, the fumigated areas were eventually reattacked by the pathogen causing the next fruit production not possible.

Nitrogen level in the soil may also affect the severity of *Fusarium* infection as NO_3 commonly decreases the severity of the disease. In contrast, NH_4 shows the opposite effect [2]. Interestingly to date, application of N fertilization in the field and its effect on *Fusarium* wilt has not been published. Nevertheless, it was reported from in vitro and hydroponic studies that high amount of NH_4 managed to block the invasion of Foc via roots [81]. The causes of these contradicting results were not elaborated.

6.3 Cultural control

As monoculture production of susceptible cultivars is difficult in infested areas, mixed plantings have been suggested as an alternative. Mixed plantings in small-scale or subsistence agriculture, in which diverse banana cultivars are planted with different crops, often develop more moderate losses than if they had been planted in monocultures [10]. Planting of Foc-tolerant or Foc-resistant cultivars accompanied by cultural practices help to improve the disease incidence and increase the crop yield [82]. Mixed culture system involving legumes, cereals, and multipurpose trees in the banana plantation can also improve the production yield of banana and develop improved tolerance against the disease [83].

6.4 Physical control

In Southern Mindanao, Philippines, and Indonesia, heat sterilization or solarization of the soil had been suggested [84]. Rice hulls were mounded on top of an affected mat and burned, supposedly generating sufficient heat to kill the pathogen. Herbert and Marx [80] stated that solarization reduced *Fusarium* wilt disease for one cycle only when it was combined with methyl bromide fumigation showing that the impact of solarization alone was not significant. In another study in Indonesia, a 6-month delay in symptom development was reported after 10 months of solarization [85].

In 1939, Dunlap used flood fallow (a method to eradicate soilborne pathogen by flooding the land) to rejuvenate Foc-infested soil [10]. However, flooded soil was rapidly recolonized by the pathogen and became impractical due to high cost needed especially in cost of labor and machinery. Thus, this method was stopped. In contrast, this measure had been commonly used in eliminating Moko disease pathogen, *Ralstonia solanacearum*, and the burrowing nematode, *Radopholus similis*, from infested soil [10].

6.5 Quarantine, exclusion, and personnel awareness

Foc is a recalcitrant pathogen that will remain in the soil for years. Thus, in halting the spread of Foc to non-infected areas, systematic quarantine and exclusion procedures are required. Stakeholders must also be educated on the infection cycle and effects of *Fusarium* wilt on banana production. This can be executed through regional awareness programs, personal communication with farmers, congresses, and contingency trainings. For example, such programs had been discussed and conducted in Latin America and the Caribbean by Bioversity International [86].

On personal level, each personnel working in the farm must be aware of the measures that they have to take in preventing the spread of *Fusarium* wilt. Some disinfectants have been found to be effective in sterilizing the farming tools, equipment, and footwear such as Sporekill (poly dimethyl ammonium chloride), Jik (sodium hypochlorite) and Prazin agri (polymeric biquanidine hydrochloride and quaternary ammonium compound) [78].

6.6 Breeding programs

The first banana breeding program involving crossbreeding of banana plantains for *Fusarium* wilt tolerance was initiated in Trinidad in 1922. The program focused on the process of crossing, screening, testing, selecting, and identifying promising hybrids for market release [87]. Generally, the success of banana breeding programs was hindered by a few barriers such as the diploid parents were lacking good agronomic and high-quality fruit traits. Other than that, low fertility of cultivar, genetic abnormalities in the parental lines, as well as the laborious duration taken for stable integration of disease resistance trait to take place from parents into the next generations contribute to the problems [88].

6.7 Transgenic research

Increasing number of publications on the development of transgenic bananas for *Fusarium* wilt tolerance were observed in the recent years probably driven by the release of whole genome sequencing of diploid banana DH Pahang [89] (**Table 2**).

Most of the published studies only reported the transgenic plants' performance under greenhouse condition. A field trial of transgenic "Cavendish" conducted in a Foc-infested area in Northern Territory of Australia was among the very few site studies published. The 3-year trial concluded that two lines of their transgenic "Cavendish" were able to resist Foc infection. One of the lines carried a disease resistance gene, RGA2 which was derived from a TR4-resistant diploid banana, while the other line overexpressed Ced9 isolated from nematode [102, 103].

Cultivar	Foc isolate	Description	Reference
Taijiao (AAA)	Race 4	• Human lysosome was used as the donor gene • Corm slice was used as the explant. Transformation was carried out via both particle bombardment and *Agrobacterium*-mediated transformation	[90]
Rasthali (AAB)	N/A	• Chitinase and β-1,3-glucanase were transformed into tiny single meristem bud of Rasthali via co-bombardment	[91]
Rasthali (AAB)	Race 1	• β-1,3-glucanase from *Glycine max* (soybean) was introduced into *Agrobacterium tumefaciens* and later transformed into single bud of Rasthali	[92]
Pei Chiao (AAA) and Gros Michel (AAA)	Race 4	• Multiple bud clumps (MBC) were used as explants for *Agrobacterium*-mediated transformation. The genes of interest were Arabidopsis root-type ferredoxin gene (Atfd3) isolated from *Arabidopsis thaliana* and ferredoxin-like protein (pflp)	[93]
Lady Finger	Race 1 (VCG 0124/5)	• Bcl-xL, Ced-9, and Bcl-2 3' UTR genes originated from chicken, *C. elegans*, and human, respectively, were used as donor genes. The genes were integrated into the banana genome via co-cultivation of *Agrobacterium*-embryogenic cell suspension	[94]
Rasthali (AAB)	Race 1	• Petunia floral defensins, PhDef1, and PhDef2 were introduced into *A. tumefaciens* and later co-cultivated with embryogenic cell suspension cultures of Rasthali	[95]
Furenzhi (AA)	Race 4	• Endochitinase gene chi42 isolated from *Trichoderma harzianum* was integrated into banana genome via co-cultivation of *A. tumefaciens* and embryogenic cell suspension of Furenzhi banana	[96]
Rasthali (AAB)	Race 1	• Intron hairpin RNA-mediated expression of two important Foc genes, *Fusarium* transcription factor 1 (ftf1) and velvet (vel), were constructed and expressed in banana cells via co-cultivation of *Agrobacterium*-embryogenic cell suspension of Rasthali	[97]
Rasthali (AAB)	Race 1	• Three different constructs harboring endogenous cell-death related genes (*MusaDAD1*, *MusaBAG1*, and *MusaBl1*) were prepared and introduced into *A. tumefaciens*. Transformation was performed via co-cultivation of *A. tumefaciens* culture and embryogenic cell suspension of Rasthali	[98]
Rasthali (AAB)	Race 1	• Seed defensin gene (Sm-AMP-D1) isolated from *Stellaria media* (common chickweed) acted as donor gene • Gene of interest was integrated via co-cultivation of *A. tumefaciens* and embryogenic cell suspension of Rasthali	[99]
Rasthali (AAB)	Race 1	• Antimicrobial peptide (*Ace-AMP1*) derived from onion seeds was used as donor gene • *Ace-AMP1* was cloned into pCAMBIA2301 and later introduced into *A. tumefaciens* • Embryogenic cell suspension of cultivar Rasthali was developed using male floral meristems and used as explant	[100]

Cultivar	Foc isolate	Description	Reference
Nangka (AAB)	Race 4	• Thaumatin-like protein (tlp) gene derived from *Oryza sativa* (rice) was used as donor gene • Particle bombardment was performed on cauliflower-like bodies of cultivar Nangka	[101]

Table 2.
Transgenic research for development of Fusarium-tolerant bananas.

Performing field trials present several difficulties which might explain the limited number of field evaluations performed on the newly developed cultivars. Among the most apparent bottleneck is the arduous process in generating and selecting the most promising lines which involved several stages starting from laboratory, greenhouse evaluation, and lastly field assessment. One has to regenerate as many lines in the laboratory since each stage would eliminate a number of unsuccessful lines leaving with only the most performing candidates for field study. The researchers may also face difficulties in finding a suitable trial site, obtaining the regulator's approval for biosafety clearance as well as convincing the locals on the project. Based on personal communication, some plantation owners refused to cooperate because they assume that the research will harm their plants and affect the gross yield. Even when the trial has commenced, researchers may still face unexpected event that could bring the trial to a halt. For example, [102, 103] initially planned for a 5-year trial to evaluate the performance of their transgenic bananas against *Fusarium* infection. Unsuccessfully, the trial was only carried out for 3 years due to invasion of another disease.

6.8 Somaclonal variation

Somaclonal variation is defined as genetic variability undergone by tissue-cultured generated plantlets [104]. Somaclonal variation instigated by in vitro micropropagation is commonly known as tissue-induced variation [105], and it occurs as a result of strong stress experienced by the in vitro-cultured plantlets [106]. Factors that contribute to somaclonal variation had been well-reviewed over the past few decades [104, 106]. The possible causes leading to this phenomenon were observed in various plant species [107–109]. However, due to complex mechanism and high degree of variation in the results, no definite conclusion was ever made [106]. Despite all of these, Larkin and Scowcroft [106] believed that chromosomal variation plays a bigger role in generating variants among the somaclones, rather than a simple base mutation. For instance, change in ploidy level, or also known as karyotypic alteration, was among the early hypothesis posed in discussing causes of somaclonal variation at chromosomal level.

The emergence of selecting *Fusarium* wilt resistant banana clones via somaclonal variation is believed to begin in Taiwan. Initiated by Taiwan Banana Research Institute (TBRI) in 1984, the institution managed to commercialize a few resistant cultivars selected from somaclonal variants in continuous greenhouse and field trials. Starting with about 20,000 tissue-cultured cv. "Cavendish" plantlets planted in Foc-infested soil, multiple screening and selection process eliminated most of the susceptible clones displaying yellowing and wilting symptoms. Rhizomes of surviving clones were collected, trimmed, and replanted again under disease condition before six final putative resistant candidates were short-listed. Among those, GCTCV-215-1 showed the most potential and was further tested on Foc-infested plantations for its disease survivability in several places in 1990. Positive results of the first trial incited the farmers to request the Taiwanese government to officially allow the growing of GTCV-215-1 for commercial purpose, even though the variety was not yet registered at that time. With only 17.2 and 5.2% disease incidence for plantlet-originating and suckers-originating, respectively, the new variety was probably nothing but a ray of hope to the growers who have been distressed with heavy loss in production. Even in the second trial, the clone continued to exhibit consistent result with only 4.8% disease incidence in comparison with 39.1% of parental Giant Cavendish. Promising field trials armed with positive response from both local and Japanese market led to the registration of GCTCV-215-1 as a commercial variety called Tai Chiao No. 1 in 1992 [110, 111].

Tai Chiao No. 1, however, was inferior to the parental Giant Cavendish in certain areas. The variety had weaker defense against wind/typhoon due to its taller and slender stature besides lower fruit bunch production and longer maturation period [110, 112]. Another study concerning somaclonal variation was carried out in Malaysia using micropropagated plantlets of "Rastali" cultivar. In the first round of trial conducted in 1994, micropropagated cv. "Rastali" showed 51% survival rate after being planted on Foc-infested soil for about 12 months. Nevertheless, a number of these plants developed split pseudostem, which is one of the Foc symptoms. Plants showing vigorous growth and did not produce split pseudostem were further selected, micropropagated, and tested again in 1996. The second trial produced promising results proving that the selected cv. "Rastali" exhibited high tolerance against *Fusarium* infection. The selected clone was later named as "Mutiara" following two successful harvests in a Foc-infested plantation [113]. Somaclonal variation approach was also used in screening *Fusarium*-resistant clones of other varieties such as Novaria and Berangan [113].

Regardless of its simplicity and feasibility, selecting a resistant clone from somaclonal variants posed several disadvantages [110]. Since it is a completely random process, a huge number of samples are required for screening. Uncontrollable frequency of successful clone and occurrence of epigenetic and off-type variants are also associated with somaclonal variation. It is commonly known that by gaining a desired trait, the plant might be losing some others. This was also demonstrated in banana [112]. Hwang and Tang [111] observed some poorer traits in resistant somaclone that included longer maturation period, lower fruit production, as well as reduced fruit quality.

6.9 Induced mutation

Induced mutation is another way of generating genetic variability in banana genomes. Working in a more efficient manner than spontaneous mutation, induced mutation happens at a higher frequency than the former [114]. Mutation breeding will increase the chances of genetic alteration by annulling the dominant allele and thus reviving the recessive allele [114]. Similar to somaclonal variation, induced mutation procedures required a large number of samples which at one time was the limiting factor to the execution of this technique [115]. To date, studies have employed two methods of mutagenizing genes, which are chemical and physical mutagenesis.

6.9.1 Chemical mutagenesis

In general, very few studies on the development of Foc-resistant banana through mutagenesis had been reported. Even though ethyl methanesulfonate (EMS) is one of the most common chemical mutagens used in studies involving plants [116–119], Bhagwat and Duncan [120] tested other two mutagens which were sodium azide

(NaN_3) and diethyl sulfate (DES). They recommended three different dosage treatments depending on the type of mutagens used. The recommended treatments for NaN_3, DES, and EMS were 30 min/2.3 mM, 60 min/20 mM, and 30 min/200 mM, respectively. When planted in greenhouse, 95.5% of the micropropagated clones showed infection symptoms and were eradicated. A total of 48 plants proceeded into field evaluation. There were 20, 9, and 19 plants coming from NaN_3, DES, and EMS treatment, respectively [120].

6.9.2 Physical mutagenesis

In vitro mutation can also be instigated by physical means such as gamma rays [116, 121–128]. Gamma rays are preferred due to their high penetration and high energy characteristics. Secondary radioactivity does not occur in gamma irradiation, and this makes the latter as a better choice over other physical mutagens like neutron [129]. Parameters used to measure radiation sensitivity and postirradiation recovery include survival rate, propagation, shoot height, and fresh weight which are all expressed as percentage of the control [130]. References [121, 130] used the same source of gamma irradiation, which was ^{60}Co, but with different experimental methodologies. According to Roux and Wingfield [130] irradiated seven banana clones of dissimilar ploidy level and genomic institution (Calcutta4 (AA), Kamaramasenge (AB), Tani (BB), Grande Naine (AAA), Williams (AAA), Three Hand Planty (AAB), and Cachaco (ABB)) with 10 doses of ^{60}Co ranging from 10 to 100 Gy at a dose rate of 44 Gy/min. The optimal dosage for diploid cultivars Calcutta4 (AA) and Tani (BB) was in the same range, which was 10–20 Gy of gamma irradiation. However, there were two different optimal dosages recommended for triploid clones. Triploid Cachaco (ABB) required the highest range of optimal dosage, which was 40–50 Gy, while triploids Three Hand Planty (AAB), Grande Naine (AAA), Williams (AAA), and Kamaramasenge (AB) were best treated with a lower dosage treatment of 30–40 Gy. Even though Kamaramasenge is a diploid, the cultivar has more triploid-like characteristics [130]. From the results, it was evident that lower dosage treatment was preferred as it could minimize the chromosome distortion and other harmful outcomes. Bhagwat and Duncan [121] in their experiment irradiated three different explants of Highgate cultivar with ^{60}Co gamma rays. All three types of explants, Type I dissected apices, Type II 4-week-old cultured apices, and Type III in vitro-cultured corms, were treated with different dose rates, which were 101.82, 177.37, and 256.8 rads/min, respectively. All explants were inoculated twice with *Fo* suspension, uprooted and re-cultured. The plants were observed for disease symptoms. Symptomless plants were chosen and replanted. Another examination was carried out 4 weeks later, and those with no infection symptoms were regarded as resistant to *Fusarium* wilt. Twenty clones were chosen for field evaluation but no further observation was reported.

7. Future strategies and opportunities

A comprehensive understanding on the roles of molecules in a cell is essential in driving the efforts for crop improvements. By taking advantage of systems biology approach, omics technologies extend the knowledge on complex interactions between genes, proteins, and metabolites of a particular species [131]. In short, omics technologies provide deeper insight on the modifications happening in an organism instigated by both internal and external factors. This may include changes in genetics, nutritional condition, and environment of an organism [132]. In association with the study of bioinformatics, different areas that are steered by omics technologies include genomics, transcriptomics, proteomics, metabolomics, and phenomics [133].

Genomics, which refer to the study and analysis on gene sequences of a species, assist researchers in specifying their genes of interest for the development of plants with improved agricultural traits. With the first banana genome sequence of a double haploid *M. acuminata* spp. *malaccensis* published in 2012 [89], endogenous gene modification approaches including gene editing technologies started to emerge. A review done by Dale et al. [103] had listed a few ongoing researches on genetically modified bananas especially on "Cavendish" banana in response to FocTR4. In addition, new technologies are being invented including RNA interference (RNAi) and gene editing tools such as CRISPR-Cas for the production of bananas with improved disease resistance, better fruit yields, and higher nutritional values. However, genomics only provide general information of the genes in a particular species [131]. In contrast, transcriptomics data reveal the expression patterns of those genes under different conditions such as biotic and abiotic stresses [131]. Current trends on transcriptomic approaches include microarrays and the next-generation sequencing (NGS) [133]. Another option suggested by [134] is the RNA-Seq on transcriptome sequencing which had been adapted by [135]. Guo et al. [135] studied about the transcriptome analysis between Foc1 and Foc4 at 48 hour postinoculation on banana variety "Brazil." It was believed that the gene contents and transcriptional regulations between Foc1 and Foc4 will contribute toward improving the banana's resistance against *Fusarium* wilt in the future [135].

Another branch of omics that has been explored is the study on proteins known as proteomics. Genomics and transcriptomics analysis generate a lot of useful data. However, the approaches do not explicitly reflect the protein expression patterns. This may be due to mRNA lifetime and the presence of non-translated RNAs [133]. Translated proteins may experience posttranslational modifications (PTMs) which are one of the limitations in genomics and transcriptomics approaches [133]. Thus, proteomics take over the task to underpin the roles of a protein [136] by analyzing the alterations that took place within the cells via protein pathways. Two-dimensional electrophoresis (2-DE) and mass spectroscopy (MS) [137] are among the approaches used widely in proteomics. Particularly for Foc, [138] performed proteomic analysis by using isobaric tags for relative and absolute quantitation (i-TRAQ)-based comparative proteomic approach on conidial germination of FocTR4. Upregulated proteins identified in the ergosterol biosynthesis pathway will be useful targets in designing an effective fungicide to tackle FocTR4 [138].

Metabolomics studies involve chemical processes that link the genotypes and phenotypes [131]. This is because alteration in both transcriptome and proteome will result in changes in metabolome as metabolome is the final downstream product of gene transcription [139]. Compared to other omics, metabolomics analysis is more complicated despite having the least domains (about 5000 metabolites) as metabolome involves varied biological components [140]. In addition, metabolic profiling presents an instantaneous picture on the current activities happening in the cell which aid in greater understanding on the mode of action of pesticide [131]. Mass spectrometry (MS) and nuclear magnetic resonance (NMR) analyses used in metabolic profiling reveal more about the metabolite regulations and changes in response to different environmental conditions. The MS approach had been adapted by Li et al. [141] to study the metabolic changes, transcriptional regulation, and signaling compounds during early stage of FocTR4 infection. Despite its importance in system biology, metabolomics approach is still not widely used [133].

By providing enormous genetic information to the researches, current technologies help to escalate the crop improvement studies. However, development of new cultivars also comes with extensive phenotypic evaluations [142]. Moreover, association of genotype to phenotype established in model systems might be deceitful [143]. In addressing this issue, plant phenomics approaches provide a noninvasive technology to study traits such as growth, performance, and composition of plants [142]. Ideally, it is hoped that the link between genotype and phenotype can eventually be described.

Unlike genotyping analysis with DNA-based molecular markers, which have been widely applied in breeding strategies, sensor-based, automated, or semiautomated phenotyping is currently underway causing delay in the plant assessment study [143]. In the study of plant pathology, several examples of sensor-based phenomics approaches include chlorophyll fluorescence imaging (CFI), multi- and hyperspectral imaging, infrared thermography (IRT), and magnetic resonance imaging (MRI) [142]. Phenomics techniques may aid in unraveling the causes of a certain plant disease as well as allowing early detection of disease-related changes in plants [142].

Omics strategies indeed provide us with gigantic amounts of data. However, the data must be validated to minimize the possibility of getting false-positive results which may affect the whole study [140]. Moving forward, it is expected that the combination between omics technology and plant breeding will generate a huge impact on crop improvement [144]. From the review done by Zhang et al. [145], multi-"omics" analyses will be the next level of omics approaches as it is able to provide more discrete and testable biological hypotheses from a large scale of high-throughput datasets. With integrated approaches, a more complex process at different levels can be analyzed, resulting in new insights to produce a more resistant species [145].

8. Conclusions

Fighting the spread of *Fusarium* wilt is a race against time. Once an area is invaded by Foc, it is almost impossible to use the area again for growing bananas. Thus, preventing the spread of *Fusarium* infection is very critical in ensuring the continuity of banana production as well as securing the nutritional supply to the consumers. In curbing the epidemics, knowledge on the etiology of *Fusarium* wilt and its infection process must be provided to the farmers and plant breeders. In addition to the current management strategies, various recent technologies introduced may shed some light into the production of *Fusarium*-resistant banana varieties thus finally putting a stop to this longstanding threat.

Acknowledgements

This work was supported by Putra Grant (Inisiatif Putra Muda) (GP-IPM/2019/9672100) financed by Universiti Putra Malaysia and Fundamental Research Grant Scheme (FRGS) (FRGS/1/2019/STG05/UPM/02/8) financed by the Ministry of Education Malaysia.

Conflict of interest

The authors declare that no conflict of interest exists.

Author details

Fatin Nadiah Jamil[1], Chu-Nie Tang[2], Noor Baity Saidi[1,2], Kok-Song Lai[2,3] and Nadiya Akmal Baharum[2*]

1 Institute of Bioscience, Universiti Putra Malaysia, Serdang, Selangor, Malaysia

2 Faculty of Biotechnology and Biomolecular Sciences, Department of Cell and Molecular Biology, Universiti Putra Malaysia, Serdang, Selangor, Malaysia

3 Health Sciences Division, Abu Dhabi Women's College, Higher Colleges of Technology, Abu Dhabi, United Arab Emirates

*Address all correspondence to: nadiya_baharum@upm.edu.my

References

[1] Ploetz RC, Pegg KG. Fungal disease of the root, corm, and pseudostem: *Fusarium* wilt. In: Jones DR, editor. Diseases of Banana, Abaca and Enset. Wallingford, UK: CAB International; 2000. pp. 143-159

[2] Ploetz RC. Management of *Fusarium* wilt of banana: A review with special reference to tropical race 4. Crop Protection. 2015;**73**:7-15. DOI: 10.1016/j.cropro.2015.01.007

[3] Siamak SB, Zheng S. Banana *Fusarium* wilt (*Fusarium oxysporum* f. sp. *cubense*) control and resistance, in the context of developing wilt-resistant bananas within sustainable production systems. Horticultural Plant Journal. 2018;**4**(5):208-218. DOI: 10.1016/j.hpj.2018.08.001

[4] Butler D. Fungus threatens top banana. Nature News. 2013;**504**(7479):195. DOI: 10.1038/504195a

[5] Moore NY, Bentley S, Pegg KG, Jones DR. Musa Disease Fact Sheet No 5. *Fusarium* Wilt of Banana. Montpellier, France: INIBAP; 1995

[6] García-Bastidas F, Ordóñez N, Konkol J, Al-Qasim M, Naser Z, Abdelwali M, et al. First report of *Fusarium oxysporum* f. sp. *cubense* tropical race 4 associated with Panama disease of banana outside Southeast Asia. Plant Disease. 2014;**98**(5):694. DOI: 10.1094/PDIS-09-13-0954-PDN

[7] Ordonez N, García-Bastidas F, Laghari HB, Akkary MY, Harfouche EN, Al Awar BN, et al. First report of *Fusarium oxysporum* f. sp. *cubense* tropical race 4 causing Panama disease in Cavendish bananas in Pakistan and Lebanon. Plant Disease. 2016;**100**(1):209. DOI: 10.1094/PDIS-12-14-1356-PDN

[8] Zheng SJ, García-Bastidas FA, Li X, Zeng L, Bai T, Xu S, et al. New geographical insights of the latest expansion of *Fusarium oxysporum* f. sp. *cubense* tropical race 4 into the greater Mekong subregion. Frontiers in Plant Science. 2018;**9**:457. DOI: 10.3389/fpls.2018.00457

[9] Bancroft J. Report of the board appointed to enquire into the cause of disease affecting livestock and plants. Queensland. In: Votes and Proceedings of Legislative Assembly 1877; 1876. p. 1011-1038

[10] Stover RH. Fusarial Wilt (Panama Disease) of Banana and Other *Musa* Species. Kew, Surrey, United Kingdom: Commonwealth Mycological Institute; 1962. 117p

[11] Ploetz RC. Diseases of tropical perennial crops: Challenging problems in diverse environments. Plant Disease. 2007;**91**(6):644-663. DOI: 10.1094/PDIS-91-6-0644

[12] Ploetz RC, Pegg KG. *Fusarium* wilt of banana and Wallace's line: Was the disease originally restricted to his indo-Malayan region? Australasian Plant Pathology. 1997;**26**(4):239-249. DOI: 10.1071/AP97039

[13] Vakili NG. *Fusarium* wilt resistance in seedlings and mature plants of *Musa* species. Phytopathology. 1965;**55**(2):135-140

[14] Pérez-vicente L, Dita MA, Martínez-de La parte E. Technical manual: Prevention and diagnostic of *Fusarium* Wilt (Panama disease) of banana caused by *Fusarium oxysporum* f. sp. *cubense* tropical race 4 (TR4). Workshop Diagnosis *Fusarium* Wilt. 2014;**4**:1-74

[15] Ploetz RC. *Fusarium* wilt of banana (Panama disease). In: Mukhopadhyay AN, Chaube HS, Kumar J, Singh US, editors. Plant

Diseases of International Importance. Vol. III. Englewood Cliffs: Pretence Hall; 1992. pp. 270-282

[16] Mostert D, Molina AB, Daniells J, Fourie G, Hermanto C, Chao CP, et al. The distribution and host range of the banana *Fusarium* wilt fungus, *Fusarium oxysporum* f. sp. *cubense*, in Asia. PLoS One. 2017;**12**(7):e0181630. DOI: 10.1371/journal.pone.0181630

[17] Ploetz RC. *Fusarium* wilt of banana. Phytopathology. 2015;**105**:1512-1521. DOI: 10.1094/PHYTO-04-15-0101-RVW

[18] Ordonez N, Seidl MF, Waalwijk C, Drenth A, Kilian A, Thomma BP, et al. Worse comes to worst: Bananas and Panama disease—When plant and pathogen clones meet. PLoS Pathogens. 2015;**11**(11):e1005197. DOI: 10.1371/journal.ppat.1005197

[19] Qi YX, Zhang X, Pu JJ, Xie YX, Zhang HQ, Huang SL. Race 4 identification of *Fusarium oxysporum* f. sp. *cubense* from Cavendish cultivars in Hainan province, China. Australasian Plant Disease Notes. 2008;**3**(1):46-47. DOI: 10.1007/BF03211234

[20] Thangavelu R. Status report on Fusarium wilt disease in India. In: 2016 ISHAS ProMusa Symposium; Montpellier, France; 2016

[21] Cook DC, Taylor AS, Meldrum RA, Drenth A. Potential economic impact of Panama disease (tropical race 4) on the Australian banana industry. Journal of Plant Diseases and Protection. 2015;**122**(5-6):229-237. DOI: 10.1007/BF03356557

[22] O'Neill WT, Henderson J, Pattemore JA, O'Dwyer C, Perry S, Beasley DR, et al. Detection of *Fusarium oxysporum* f. sp. *cubense* tropical race 4 strain in northern Queensland. Australasian Plant Disease Notes. 2016;**11**(1):33. DOI: 10.1007/s13314-016-0218-1

[23] Hung TN, Hung NQ, Mostert D, Viljoen A, Chao CP, Molina AB. First report of *Fusarium* wilt on Cavendish bananas, caused by *Fusarium oxysporum* f. sp. *cubense* tropical race 4 (VCG 01213/16), in Vietnam. Plant Disease. 2018;**102**(2):448. DOI: 10.1094/PDIS-08-17-1140-PDN

[24] Chittarath K, Mostert D, Crew KS, Viljoen A, Kong G, Molina AB, et al. First report of *Fusarium oxysporum* f. sp. *cubense* tropical race 4 (VCG 01213/16) associated with Cavendish bananas in Laos. Plant Disease. 2018;**102**(2):449. DOI: 10.1094/PDIS-08-17-1197-PDN

[25] Maymon M, Shpatz U, Harel YM, Levy E, Elkind G, Teverovsky E, et al. First report of *Fusarium oxysporum* f. sp. *cubense* tropical race 4 causing *Fusarium* wilt of Cavendish bananas in Israel. Plant Disease. 2018;**102**(12):2655. DOI: 10.1094/PDIS-05-18-0822-PDN

[26] Lambert J. Alarm as Devastating Banana Fungus Reaches the Americas. Nature News. 19 August 2019. ISSN: 1476-468. Available online: https://www.nature.com/articles/d41586-019-02489-5

[27] Dong X, Ling N, Wang M, Shen Q, Guo S. Fusaric acid is a crucial factor in the disturbance of leaf water imbalance in *Fusarium*-infected banana plants. Plant Physiology and Biochemistry. 2012;**60**:171-179. DOI: 10.1016/j.plaphy.2012.08.004

[28] Li C, Yang J, Li W, Sun J, Peng M. Direct root penetration and rhizome vascular colonization by *Fusarium oxysporum* f. sp. *cubense* are the key steps in the successful infection of Brazil Cavendish. Plant Disease. 2017;**101**(12):2073-2078. DOI: 10.1094/PDIS-04-17-0467-RE

[29] Guo L, Yang L, Liang C, Wang G, Dai Q, Huang J. Differential colonization patterns of bananas (Musa spp.) by physiological race 1 and race

4 isolates of *Fusarium oxysporum* f. sp. *cubense*. Journal of Phytopathology. 2015;**163**(10):807-817. DOI: 10.1111/jph.12378

[30] Li C, Chen S, Zuo C, Sun Q, Ye Q, Yi G, et al. The use of GFP-transformed isolates to study infection of banana with *Fusarium oxysporum* f. sp. *cubense* race 4. European Journal of Plant Pathology. 2011;**131**(2):327-340. DOI: 10.1007/s10658-011-9811-5

[31] Trujillo EE. Pathological-anatomical studies of Gros Michel banana affected by *Fusarium* wilt. Phytopathology. 1963;**53**(2):162

[32] Beckman CH. The Nature of Wilt Diseases of Plants. St Paul: America Phytopathological Press; 1987

[33] Beckman CH. Host responses to the pathogen. In: Ploetz RC, editor. Fusarium Wilt of Banana. USA: APS Press; 1990. pp. 93-105

[34] Ho YN, Chiang HM, Chao CP, Su CC, Hsu HF, Guo CT, et al. In planta biocontrol of soilborne *Fusarium* wilt of banana through a plant endophytic bacterium, *Burkholderia cenocepacia* 869T2. Plant and Soil. 2015;**387**(1-2):295-306. DOI: 10.1007/s11104-014-2297-0

[35] Zhang L, Yuan T, Wang Y, Zhang D, Bai T, Xu S, et al. Identification and evaluation of resistance to *Fusarium oxysporum* f. sp. *cubense* tropical race 4 in *Musa acuminata* Pahang. Euphytica. 2018;**214**(7):106. DOI: 10.1007/s10681-018-2185-4

[36] Stover RH. Studies on *Fusarium* wilt of bananas: IX. Competitive saprophytic ability of *F. oxysporum* f. *cubense*. Canadian Journal of Botany. 1962;**40**(11):1473-1481

[37] Simmonds NW. Bananas. 2nd ed. London: Longmans; 1966. p. 512

[38] Buddenhagen I. Understanding strain diversity in *Fusarium oxysporum* f. sp. *cubense* and history of introduction of 'tropical race 4' to better manage banana production. Acta Horticulturae. 2009;**828**:193-204. DOI: 10.17660/ActaHortic.2009.828.19

[39] Bentley S, Pegg KG, Dale JL. Genetic variation among a world-wide collection of isolates of *Fusarium oxysporum* f. sp. *cubense* analysed by RAPD-PCR fingerprinting. Mycological Research. 1995;**99**(11):1378-1384. DOI: 10.1016/S0953-7562(09)81225-2

[40] Bentley S, Pegg KG, Moore NY, Davis RD, Buddenhagen IW. Genetic variation among vegetative compatibility groups of *Fusarium oxysporum* f. sp. *cubense* analyzed by DNA fingerprinting. Phytopathology. 1998;**88**(12):1283-1293. DOI: 10.1094/PHYTO.1998.88.12.1283

[41] Boehm EW, Ploetz RC, Kistler HC. Statistical analysis of electrophoretic karyotype variation among vegetative compatibility groups of *Fusarium oxysporum* f. sp. *cubense*. Molecular Plant-Microbe Interactions. 1994;**7**(2):196-207

[42] Fourie G, Steenkamp ET, Ploetz RC, Gordon TR, Viljoen A. Biology and evolution of *Fusarium oxysporum*, with specific reference to the banana wilt pathogen. Infection, Genetics and Evolution. 2011;**11**:533-542

[43] Koenig RL, Ploetz RC, Kistler HC. *Fusarium oxysporum* f. sp. *cubense* consists of a small number of divergent and globally distributed clonal lineages. Phytopathology. 1997;**87**(9):915-923. DOI: 10.1094/PHYTO.1997.87.9.915

[44] O'Donnell K, Kistler HC, Cigelnik E, Ploetz RC. Multiple evolutionary origins of the fungus causing Panama disease of banana: Concordant evidence from nuclear and mitochondrial gene

genealogies. Proceedings of the National Academy of Sciences. 1998;**95**(5):2044-2049. DOI: 10.1073/pnas.95.5.2044

[45] Islands C, Province M. Vegetative compatibility among races of *Fusarium oxysporum* f. sp. *cubense*. Plant Disease. 1988;**72**:325-328

[46] Stover RH. Fusarium wilt of banana: some history and current status of the disease. In: Ploetz RC, editor. Fusarium wilt of banana. Minnesota: American Phytopathological Society; 1990. p. 1-7

[47] Stover RH, Buddenhagen IW. Banana breeding: Polyploidy, disease resistance and productivity. Fruits. 1986;**41**(3):175-214

[48] Stover RH, Simmonds NW. Bananas. Tropical Agricultural Series. Vol. 10. New York: John Wiley and Sons, Inc.; 1987. pp. 158-468

[49] Waite BH. Wilt of *Heliconia* spp. caused by *Fusarium oxysporum* f. sp. *cubense* race 3. Tropical Agriculture Trinidad. 1963;**40**:299-305

[50] Leong SK, Latiffah Z, Baharuddin S. Genetic diversity of *Fusarium oxysporum* f. sp. *cubense* isolates from Malaysia. African Journal of Microbiology Research. 2010;**4**(11):1026-1037

[51] Mayers PE. *Fusarium* wilt of Cavendish bananas in Queensland. In: Liddell C, editor. Proceedings of the International *Fusarium* Workshop. Australia: The University of Syndey 5; 1983. p. 60

[52] Su HJ, Hwang SC, Ko WH. Fusarial wilt of Cavendish bananas in Taiwan. Plant Disease. 1986;**70**(9):814-818

[53] Ploetz RC. *Fusarium* wilt of banana is caused by several pathogens referred to as *Fusarium oxysporum* f. sp. *cubense*. Phytopathology. 2006;**96**(6):653-656. DOI: 10.1094/PHYTO-96-0653

[54] Thangavelu R, Mustaffa MM. Current advances in the Fusarium wilt disease management in banana with emphasis on biological control. In: Cumagun CJ, editor. Plant Pathology. Rijeka: Intech; 2012. pp. 273-298

[55] Ploetz RC. Panama disease: An old nemesis rears its ugly head: Part 1. The beginnings of the banana export trades. Plant Health Progress. 2005;**6**(1):18. DOI: 10.1094/PHP-2005-1221-01-RV

[56] Snyder WC, Hansen HN. The species concept in *Fusarium*. American Journal of Botany. 1940;**27**(2):64-67. DOI: 10.1002/j.1537-2197.1940.tb14217.x

[57] Morán JF. Improvement of Cavendish banana cultivars through conventional breeding. Acta Horticulturae. 2013;**986**:205-208. DOI: 10.17660/ActaHortic.2013.986.21

[58] Fravel D, Olivain C, Alabouvette C. *Fusarium oxysporum* and its biocontrol. The New Phytologist. 2003;**157**(3):493-502. DOI: 10.1046/j.1469-8137.2003.00700.x

[59] Jie L, Zifeng W, Lixiang C, Hongming T, Patrik I, Zide J, et al. Artificial inoculation of banana tissue culture plantlets with indigenous endophytes originally derived from native banana plants. Biological Control. 2009;**51**(3):427-434. DOI: 10.1016/j.biocontrol.2009.08.002

[60] Gang G, Bizun W, Weihong M, Xiaofen L, Xiaolin Y, Chaohua Z, et al. Biocontrol of *Fusarium* wilt of banana: Key influence factors and strategies. African Journal of Microbiology Research. 2013;**7**(41):4835-4843. DOI: 10.5897/AJMR2012.2392

[61] Raza W, Ling N, Zhang R, Huang Q, Xu Y, Shen Q. Success evaluation of

the biological control of *Fusarium* wilts of cucumber, banana, and tomato since 2000 and future research strategies. Critical Reviews in Biotechnology. 2017;**37**(2):202-212. DOI: 10.3109/07388551.2015.1130683

[62] Akila R, Rajendran L, Harish S, Saveetha K, Raguchander T, Samiyappan R. Combined application of botanical formulations and biocontrol agents for the management of *Fusarium oxysporum* f. sp. *cubense* (*Foc*) causing *Fusarium* wilt in banana. Biological Control. 2011;**57**(3):175-183. DOI: 10.1016/j.biocontrol.2011.02.010

[63] Getha K, Vikineswary S, Wong WH, Seki T, Ward A, Goodfellow M. Evaluation of *Streptomyces* sp. strain g10 for suppression of *Fusarium* wilt and rhizosphere colonization in pot-grown banana plantlets. Journal of Industrial Microbiology and Biotechnology. 2005;**32**(1):24-32. DOI: 10.1007/s10295-004-0199-5

[64] Getha K, Vikineswary S. Antagonistic effects of *Streptomyces violaceusniger* strain G10 on *Fusarium oxysporum* f. sp. *cubense* race 4: Indirect evidence for the role of antibiosis in the antagonistic process. Journal of Industrial Microbiology and Biotechnology. 2002;**28**(6):303-310. DOI: 10.1038/sj/jim/7000247

[65] Thangavelu R, Mustaffa MM. First report on the occurrence of a virulent strain of *Fusarium* wilt pathogen (Race-1) infecting Cavendish (AAA) group of bananas in India. Plant Disease. 2010;**94**(11):1379. DOI: 10.1094/PDIS-05-10-0330

[66] Ting AS, Meon S, Kadir J, Radu S, Singh G. Endophytic microorganisms as potential growth promoters of banana. BioControl. 2008;**53**(3):541-553. DOI: 10.1007/s10526-007-9093-1

[67] Ting AS, Mah SW, Tee CS. Prevalence of endophytes antagonistic towards *Fusarium oxysporum* f. sp. *cubense* race 4 in various plants. American-Eurasian Journal of Sustainable Agriculture. 2009;**3**(3):399-406

[68] Ting AS, Mah SW, Tee CS. Detection of potential volatile inhibitory compounds produced by endobacteria with biocontrol properties towards *Fusarium oxysporum* f. sp. *cubense* race 4. World Journal of Microbiology and Biotechnology. 2011;**27**(2):229-235. DOI: 10.1007/s11274-010-0447-y

[69] Thangavelu R, Palaniswami A, Velazhahan R. Mass production of *Trichoderma harzianum* for managing *Fusarium* wilt of banana. Agriculture, Ecosystems and Environment. 2004;**103**(1):259-263. DOI: 10.1016/j.agee.2003.09.026

[70] Mohammed AM, AL-Ani LK, Bekbayeva L, Salleh B. Biological control of *Fusarium oxysporum* f. sp. *cubense* by *Pseudomonas fluorescens* and BABA *in vitro*. World Applied Sciences Journal. 2011;**15**(2):189-191

[71] Gnanasekaran P, Salique SM, Panneerselvam A, Uma Maheswari K. *In vitro* biological control of *Fusarium oxysporum* f. sp. *cubense* by using some Indian medicinal plants. International Journal of Current Research and Academic Review. 2015;**3**:107-116

[72] Sekhar AC, Thomas P. Isolation and identification of shoot-tip associated endophytic bacteria from banana cv. Grand Naine and testing for antagonistic activity against *Fusarium oxysporum* f. sp. *cubense*. American Journal of Plant Sciences. 2015;**6**(07):943-954. DOI: 10.4236/ajps.2015.67101

[73] Ali SM, Yousef NM, Nafady NA. Application of biosynthesized silver nanoparticles for the control of land snail *Eobania vermiculata* and some plant pathogenic fungi. Journal of Nanomaterials. 2015;**2015**:3. DOI: 10.1155/2015/218904

[74] Köberl M, Dita M, Martinuz A, Staver C, Berg G. Members of Gammaproteobacteria as indicator species of healthy banana plants on *Fusarium* wilt-infested fields in Central America. Scientific Reports. 2017;**7**(45318):1-9. DOI: 10.1038/srep45318

[75] Ploetz RC. Biological control of *Fusarium* wilt: A review and an evaluation. In: Abstract Booklet, International Congress on Musa: Harnessing Research to Improve Livelihoods. International Network for the Improvement of Banana and Plantain (Bioversity International) Online Publication; 2004. p. 141

[76] Thangavelu R, Jayanthi A. RFLP analysis of rDNA-ITS regions of native non-pathogenic *Fusarium oxysporum* isolates and their field evaluation for the suppression of *Fusarium* wilt disease of banana. Australasian Plant Pathology. 2009;**38**(1):13-21. DOI: 10.1071/AP08071

[77] Belgrove A, Steinberg C, Viljoen A. Evaluation of nonpathogenic *Fusarium oxysporum* and *Pseudomonas fluorescens* for Panama disease control. Plant Disease. 2011;**95**(8):951-959. DOI: 10.1094/PDIS-06-10-0409

[78] Nel B, Steinberg C, Labuschagne N, Viljoen A. Evaluation of fungicides and sterilants for potential application in the management of *Fusarium* wilt of banana. Crop Protection. 2007;**26**(4):697-705. DOI: 10.1016/j.cropro.2006.06.008

[79] Lakshmanan P, Selvaraj P, Mohan S. Efficacy of different methods for the control of Panama disease. Tropical Pest Management. 1987;**33**(4):373-374. DOI: 10.1080/09670878709371187

[80] Herbert JA, Marx D. Short-term control of Panama disease of bananas in South Africa. Phytophylactica. 1990;**22**(3):339-340

[81] Zhang MX, Zhang MC, Chen P, Ruan Y, Zhu YY, Shen QR. Influence of nitrate/ammonium ratio on the plant growth of banana and related wilt disease development. Journal of Plant Nutrition and Fertilizer. 2013;**19**:1241-1247

[82] Haddad F, Rocha LS, Soares AC, Martins IP, Teixeira LA, Staver C, Dita M. Management of Fusarium wilt of bananas in Minas Gerais, Brazil. Acta Horticulturae 1196; 2016. pp. 137-146. DOI: 10.17660/ActaHortic.2018.1196.16

[83] Dowiya NB, Rweyemamu CL, Maerere AP. Banana (*Musa* spp. Colla) cropping systems, production constraints and cultivar preferences in eastern Democratic Republic of Congo. Journal of Animal and Plant Sciences. 2009;**4**(2):341-356

[84] Molina A, Williams R, Hermanto C, Suwanda M, Komolong B, Kokoa P. Mitigating the threat of banana *Fusarium* wilt: Understanding the agroecological distribution of pathogenic forms and developing disease management strategies. Australia: Australian Centre for International Agricultural Research, GPO Box. 2010;**1571**:2010-2012

[85] Hermanto C, Djatnika I, Emilda D. Pre-planting treatments for management of banana *Fusarium* wilt. Journal of Agricultural and Biological Science. 2012;**7**(4):260-265

[86] Pocasangre LE, Ploetz RC, Molina AB, Vicente LP. Raising awareness of the threat of *Fusarium* wilt tropical race 4 in Latin America and the Caribbean. 2011;**897**:331-337. DOI: 10.17660/ActaHortic.2011.897.45

[87] Ortiz R. Conventional banana and plantain breeding. In: Amorim EP, Pires de Matos, Ven den Gergh I, editors. VII International Symposium on Banana: ISHS-ProMusa Symposium on Bananas

and Plantains: Towards Sustainable Global Production 986; 10-14 October 2011; Salvador, Brazil, Belgium: International Society for Horticultural Science; 2013. pp. 177-194. DOI: 10.17660/ ActaHortic.2013.986.19

[88] Ortiz R, Swennen R. From crossbreeding to biotechnology-facilitated improvement of banana and plantain. Biotechnology Advances. 2014;**32**(1):158-169. DOI: 10.1016/j.biotechadv.2013.09.010

[89] D'Hont A, Denoeud F, Aury JM, Baurens FC, Carreel F, Garsmeur O, et al. The banana (*Musa acuminata*) genome and the evolution of monocotyledonous plants. Nature. 2012;**488**(7410):213-217. DOI: 10.1038/nature11241

[90] Pei XW, Chen SK, Wen RM, Ye S, Huang JQ, Zhang YQ, et al. Creation of transgenic bananas expressing human lysozyme gene for Panama wilt resistance. Journal of Integrative Plant Biology. 2005;**47**(8):971-977. DOI: 10.1111/j.1744-7909.2005.00141.x

[91] Sreeramanan S, Maziah M, Abdullah MP, Sariah M, Xavier R. Transient expression of *gusA* and *gfp* gene in *Agrobacterium*-mediated banana transformation using single tiny meristematic bud. Asian Journal of Plant Sciences. 2006;**5**(3):468-480. DOI: 10.3923/ajps.2006.468.480

[92] Maziah M, Sariah M, Sreeramanan S. Transgenic banana Rastali (AAB) with *β-1, 3-glucanase* gene for tolerance to *Fusarium* wilt race 1 disease via *Agrobacterium*-mediated transformation system. The Plant Pathology Journal. 2007;**6**:271-282. DOI: 10.1007/s11816-011-0179-y

[93] Yip MK, Lee SW, Su KC, Lin YH, Chen TY, Feng TY. An easy and efficient protocol in the production of *pflp* transgenic banana against *Fusarium* wilt. Plant Biotechnology Reports. 2011;**5**(3):245-254

[94] Paul JY, Becker DK, Dickman MB, Harding RM, Khanna HK, Dale JL. Apoptosis-related genes confer resistance to *Fusarium* wilt in transgenic 'lady finger' bananas. Plant Biotechnology Journal. 2011;**9**(9):1141-1148. DOI: 10.1111/j.1467-7652.2011.00639.x

[95] Ghag SB, Shekhawat UK, Ganapathi TR. Petunia floral defensins with unique prodomains as novel candidates for development of *Fusarium* wilt resistance in transgenic banana plants. PLoS One. 2012;**7**(6):e39557. DOI: 10.1371/journal.pone.0039557

[96] Hu CH, Wei YR, Huang YH, Yi GJ. An efficient protocol for the production of *chit42* transgenic Furenzhi banana (*Musa* spp. AA group) resistant to *Fusarium oxysporum*. In Vitro Cellular and Developmental Biology. Plant. 2013;**49**(5):584-592. DOI: 10.1007/s11627-013-9525-9

[97] Ghag SB, Shekhawat UK, Ganapathi TR. Host-induced post-transcriptional hairpin RNA-mediated gene silencing of vital fungal genes confers efficient resistance against *Fusarium* wilt in banana. Plant Biotechnology Journal. 2014;**12**(5):541-553. DOI: 10.1111/pbi.12158

[98] Ghag SB, Shekhawat UK, Ganapathi TR. Native cell-death genes as candidates for developing wilt resistance in transgenic banana plants. Annals of Botany Plants. 2014;**6**:plu1037. DOI: 10.1093/aobpla/plu037

[99] Ghag SB, Shekhawat UK, Ganapathi TR. Transgenic banana plants expressing a *Stellaria media* defensin gene (*Sm-AMP-D1*) demonstrate improved resistance to *Fusarium oxysporum*. Plant Cell, Tissue

and Organ Culture. 2014;**119**(2):247-255. DOI: 10.1007/s11240-014-0529-x

[100] Mohandas S, Sowmya HD, Saxena AK, Meenakshi S, Rani RT, Mahmood R. Transgenic banana cv. Rasthali (AAB, silk gp) harboring *Ace-AMP1* gene imparts enhanced resistance to *Fusarium oxysporum* f. sp. *cubense* race 1. Scientia Horticulturae. 2013;**164**:392-399. DOI: 10.1016/j.scienta.2013.09.018

[101] Mahdavi F, Sariah M, Maziah M. Expression of rice thaumatin-like protein gene in transgenic banana plants enhances resistance to *Fusarium* wilt. Applied Biochemistry and Biotechnology. 2012;**166**(4):1008-1019. DOI: 10.1007/s12010-011-9489-3

[102] Dale J, James A, Paul JY, Khanna H, Smith M, Peraza-Echeverria S, et al. Transgenic Cavendish bananas with resistance to *Fusarium* wilt tropical race 4. Nature Communications. 2017;**8**(1):1496. DOI: 10.1038/s41467-017-01670-6

[103] Dale J, Paul JY, Dugdale B, Harding R. Modifying bananas: From transgenics to organics? Sustainability. 2017;**9**(3):333. DOI: 10.3390/su9030333

[104] Lee M, Phillips RL. The chromosomal basis of somaclonal variation. Annual Review of Plant Physiology and Plant Molecular Biology. 1988;**39**(1):413-437

[105] Kaeppler SM, Kaeppler HF, Rhee Y. Epigenetic aspects of somaclonal variation in plants. In: Matzke MA, Matzke AJM, editors. Plant Gene Silencing. Dordrecht: Springer; 2000. pp. 59-68. DOI: 10.1007/978-94-011-4183-3_4

[106] Larkin PJ, Scowcroft WR. Somaclonal variation—A novel source of variability from cell cultures for plant improvement. Theoretical and Applied Genetics. 1981;**60**(4):197-214. DOI: 10.1007/BF02342540

[107] Krishnamurthi M, Tlaskal J. Fiji disease resistant *Saccharum officinarum* var. Pindar subclones from tissue cultures. Proceedings of International Society Sugar Cane Technologists. 1974;**15**(1):130-137

[108] Liu MC, Chen WH. Tissue and cell culture as aids to sugarcane breeding. I. Creation of genetic variation through callus culture. Euphytica. 1976;**25**(1):393-403. DOI: 10.1007/BF00041572

[109] Nishi T, Yamada Y, Takahashi E. Organ redifferentiation and plant restoration in rice callus. Nature. 1968;**219**(5153):508-509. DOI: 10.1038/219508a0

[110] Hwang SC, Ko WH. Cavendish banana cultivars resistant to *Fusarium* wilt acquired through somaclonal variation in Taiwan. Plant Disease. 2004;**88**(6):580-588. DOI: 10.1094/PDIS.2004.88.6.580

[111] Hwang SC, Tang CY. Unconventional banana breeding in Taiwan. In: Disease of Banana, Abaca and Enset. Wallingford: Centre for Agriculture and Biosciences International Publishing; 1999. 544p

[112] Tang CY. Somaclonal variation: A tool for the improvement of Cavendish banana cultivars.In: Chang CW, Drew R, editors. II International Symposium on Biotechnology of Tropical and Subtropical Species 692; 5 November 2001; Taiwan; Belgium: International Society for Horticultural Science; 2005. pp. 61-66. DOI: 10.17660/ActaHortic.2005.692.6

[113] Ho YW, Mak C, Liew KW. Selection of banana cultivars for tolerance to fusarium wilt race 4 in Malaysia.

In: Molina AB, Nik Masdek NH, Liew KW, editors. Banana *Fusarium* Wilt Management: Towards Sustainable Cultivation; International Network for the Improvement of Bananas and Plantains—Asia and the Pacific Network; Los Banos, Laguna, Philippines; 2001. pp. 234-242

[114] Jain SM, Swennen R. Banana improvement: Cellular, molecular biology, and induced mutations. In: Proceedings from a Meeting Held on 24-28 September 2001; Enfield, NH, USA: Science Publishers Inc.; 2004. p. 382

[115] Jain SM. In vitro mutagenesis in Banana (Musa spp.) improvement. In: IV International Symposium on Banana; International Conference on Banana and Plantain in Africa. Harnessing International 879; 2008. pp. 605-614. DOI: 10.17660/ActaHortic.2010.879.67

[116] Epp D. Somaclonal variation in banana: A case study with Fusarium wilt. In: Persley GJ, De Langhe EA, editors. Banana and Plantain Breeding Strategies. Canberra: Australian Centre for International Agriculture Research; 1987. pp. 140-150

[117] Kim Y, Schumaker KS, Zhu JK. EMS mutagenesis of *Arabidopsis*. In: Salinas J, Sanchez-Serrano JJ, editors. *Arabidopsis* Protocols. Methods in Molecular Biology™. Vol. 323. Totowa, NJ: Humana Press; 2006. pp. 101-103. DOI: 10.1385/1-59745-003-0:101

[118] Lai KS, Kaothien-Nakayama P, Iwano M, Takayama S. A tilling resource for functional genomics in *Arabidopsis thaliana* accession C24. Genes and Genetic Systems. 2012;**87**(5):291-297. DOI: 10.1266/ggs.87.291

[119] Omar MS, Novak FJ, Brunner H. *In vitro* action of ethylmethanesulphonate on banana shoot tips. Scientia Horticulturae. 1989;**40**(4):283-295. DOI: 10.1016/0304-4238(89)90102-7

[120] Bhagwat B, Duncan EJ. Mutation breeding of banana cv. Highgate (*Musa* spp., AAA Group) for tolerance to *Fusarium oxysporum* f. sp. *cubense* using chemical mutagens. Scientia Horticulturae. 1998;**73**(1):11-22. DOI: 10.1016/S0304-4238(97)00141-6

[121] Bhagwat B, Duncan EJ. Mutation breeding of Highgate (*Musa acuminata*, AAA) for tolerance to *Fusarium oxysporum* f. sp. *cubense* using gamma irradiation. Euphytica. 1998;**101**(2): 143-150. DOI: 10.1023/A:1018391619986

[122] Espino RR, Zamora AB, Pimentel RB. Mutation breeding on selected Philippine fruit crops. In: Nuclear Techniques and In Vitro Culture for Plant Improvement. Vienna: International Atomic Energy Agency; 1986. pp. 429-433

[123] Hautea DM, Molina GC, Balatero CH, Coronado NB, Perez EB, Alvarez MTH, et al. Analysis of induced mutants of Philippine bananas with molecular markers. In: Jain SM, Swennen R, editors. Banana Improvement: Cellular, Molecular Biology, and Induced Mutations. Enfield, NH, USA: Science Publishers Inc.; 2004. p. 382

[124] López J, Strosse H, de la C Ventura J, Sánchez R, Rodriguez S, Swennen R, Panis B, Afza R. Field evaluation of potential mutants obtained after gamma irradiation of banana and plantain (*Musa* spp.) shoot-tip and embryogenic cell cultures. In: Jain SM and Swennen R, editors. Banana Improvement: Cellular, Molecular Biology and Induced Mutations. Science Publishers Inc., Enfield, NH, USA; 2004. pp. 87-96

[125] Mak C, Ho YW, Tan YP, Rusli I, Liew KW. Mutation induction by gamma irradiation in a triploid banana Pisang Berangan. Malaysian Journal of Science Series A Life Sciences. 1995;**16**(2):77-81

[126] Novak FJ, Afza R, Phadvibulya V, Hermelin T, Brunner H, Donini B. Micropropagation and radiation sensitivity in shoot-tip cultures of banana and plantain. In: Nuclear Techniques and *In Vitro* Culture for Plant Improvement. Vienna: International Atomic Energy Agency; 1986. pp. 167-174

[127] Roux NS, Toloza A, Dolezel J, Panis B. Usefulness of embryogenic cell suspension cultures for the induction and selection of mutants in *Musa* spp. In: Mohan Jain S, Swennen R, editors. Banana Improvement: Cellular, Molecular Biology and Induced Mutations. Enfield, NH, USA: Science Publishers Inc.; 2004. pp. 33-43

[128] Smith MK, Hamill SD, Langdon PW, Pegg KG. *In vitro* mutation breeding for the development of bananas with resistance to race 4, *Fusarium* wilt. In: *In Vitro* Mutation Breeding of Banana and Plantain. Proceedings of Reports of an FAO/IAEA Coordinated Research Programme (1988-1993). Vienna, Austria: IAEA; 1995. pp. 37-44

[129] Kulkarni VM, Ganapathi TR, Suprasanna P, Bapat VA. *In vitro* mutagenesis in banana (Musa spp.) using gamma irradiation. In: Mohan JS, Haggman H, editors. Protocols for Micropropagation of Woody Trees and Fruits. The Netherlands: Springer; 2007. pp. 543-559. DOI: 10.1007/978-1-4020-6352-7_48

[130] Roux J, Wingfield MJ. Survey and virulence of fungi occurring on diseased *Acacia mearnsii* in South Africa. Forest Ecology and Management. 1997;**99**(3):327-336. DOI: 10.1016/S0378-1127(97)00110-2

[131] Van Emon JM. The omics revolution in agricultural research. Journal of Agricultural and Food Chemistry. 2015;**64**(1):36-44. DOI: 10.1021/acs.jafc.5b04515

[132] Chassy BM. Can–omics inform a food safety assessment? Regulatory Toxicology and Pharmacology. 2010;**58**(3):S62-S70. DOI: 10.1016/j.yrtph.2010.05.009

[133] Yadav D, Tanveer A, Malviya N, Yadav S. Overview and principles of bioengineering: The drivers of omics technologies. In: Barh D, Azevedo V, editors. Omics Technologies and Bio-Engineering: Volume 1: Towards Improving Quality of Life. Cambridge: Academic Press; 2018. pp. 3-23. DOI: 10.1016/B978-0-12-804659-3.00001-4

[134] Sharma CM, Vogel J. Differential RNA-seq: The approach behind and the biological insight gained. Current Opinion in Microbiology. 2014;**19**:97-105. DOI: 10.1016/j.mib.2014.06.010

[135] Guo L, Han L, Yang L, Zeng H, Fan D, Zhu Y, et al. Genome and transcriptome analysis of the fungal pathogen *Fusarium oxysporum* f. sp. *cubense* causing banana vascular wilt disease. PLoS One. 2014;**9**(4):e95543. DOI: 10.1371/journal.pone.0095543

[136] Vlahou A, Fountoulakis M. Proteomic approaches in the search for disease biomarkers. Journal of Chromatography B. 2005;**814**(1):11-19. DOI: 10.1016/j.jchromb.2004.10.024

[137] Eldakak M, Milad SI, Nawar AI, Rohila JS. Proteomics: A biotechnology tool for crop improvement. Frontiers in Plant Science. 2013;**4**:35. DOI: 10.3389/fpls.2013.00035

[138] Deng GM, Yang QS, He WD, Li CY, Yang J, Zuo CW, et al. Proteomic analysis of conidia germination in *Fusarium oxysporum* f. sp. *cubense* tropical race 4 reveals new targets in ergosterol biosynthesis pathway for controlling *Fusarium* wilt of banana. Applied Microbiology and Biotechnology. 2015;**99**(17):7189-7207. DOI: 10.1007/s00253-015-6768-x

[139] Urbanczyk-Wochniak E, Luedemann A, Kopka J, Selbig J, Roessner-Tunali U, Willmitzer L, et al. Parallel analysis of transcript and metabolic profiles: A new approach in systems biology. European Molecular Biology Organization reports. 2003;**4**(10):989-993. DOI: 10.1038/sj.embor.embor944

[140] Horgan RP, Kenny LC. SAC review: omic technologies: Genomics, transcriptomics. Proteomics and metabolomics. The Obstetrician and Gynaecologist. 2011;**13**:189-195. DOI: 10.1576/toag.13.3.189.27672

[141] Li W, Li C, Sun J, Peng M. Metabolomic, biochemical, and gene expression analyses reveal the underlying responses of resistant and susceptible Banana species during early infection with *Fusarium oxysporum* f. sp. *cubense*. Plant Disease. 2017;**101**(4):534-543. DOI: 10.1094/PDIS-09-16-1245-RE

[142] Simko I, Jimenez-Berni JA, Sirault XR. Phenomic approaches and tools for phytopathologists. Phytopathology. 2016;**107**(1):6-17. DOI: 10.1094/PHYTO-02-16-0082-RVW

[143] Furbank RT, Tester M. Phenomics–technologies to relieve the phenotyping bottleneck. Trends in Plant Science. 2011;**16**(12):635-644. DOI: 10.1016/j.tplants.2011.09.005

[144] Witzel K, Neugart S, Ruppel S, Schreiner M, Wiesner M, Baldermann S. Recent progress in the use of omics technologies in brassicaceous vegetables. Frontiers in Plant Science. 2015;**6**:244. DOI: 10.3389/fpls.2015.00244

[145] Zhang W, Li F, Nie L. Integrating multiple 'omics' analysis for microbial biology: Application and methodologies. Microbiology. 2010;**156**(2):287-301. DOI: 10.1099/mic.0.034793-0

Chapter 8

Mango Diseases: Impact of Fungicides

Muhammad Ibrahim Khaskheli

Abstract

Mango, *Mangifera indica* L., is known to be the king of all fruits due to its delicious taste, marvelous fragrance, and beautiful appearance. However, several infectious diseases caused by many phytopathogens are deteriorating mango quality and quantity. Mango tree and fruit have been affected by about 83 diseases reported worldwide, and in Pakistan, 27 diseases are recognized as more important. Disease control always remains a challenge for the farmers to get optimum production especially due to pesticide resistance. Resistance to fungicide in current days is a major threat to plant disease management. In many cases, plant pathogen resistance could develop naturally; thus, several newly developed chemistries of fungicides remain at high risk. However, research toward an increase of resistance and delay in disease development has been undertaken. Existing fungicide chemistry, sometimes, renamed with new trade name does not satisfy the farmer to apply such fungicides for disease management. However, chemical fungicides are believed to be a significant way to control fungal pathogens or sometimes to inhibit and prevent the development of pathogens. However, due to pathogen resistance development, it is hard to manage plant diseases. Therefore, the impacts of such fungicide management in some important mango diseases are discussed in this chapter.

Keywords: mango diseases, cause, management, fungicides, resistance development

1. Introduction

Mango, *Mangifera indica* L., the king of all fruits, belongs to the family Anacardiaceae and order Sapindales [1]. Generally, it is grown in tropical and subtropical regions of Southeast Asia [2]. The native home of mango is considered as India to Burma (Myanmar) or maybe from the Malay region. Since the sixteenth century, it traveled to other parts of the world [3]. Mango was introduced from India to other countries of the tropical and subtropical world mainly by the Muslim missionaries, Spanish voyagers, and Portuguese [4–6]. The great Mughal Emperors, especially Akbar, made the preliminary contribution by establishing the Lakh Bagh through the selection and subsequent cloning. In other reports, mango seed traveling begin around 300 or 400 AD from Asia to the Middle East, East Africa, and South America [7, 8]. Some mango growers believed that Malaysian region is the original home of mango, due to a maximum number of species grown over there, whereas the recorded numbers of grown species are about 20. However, in Eastern India and Burma, the cultivation of mango reported in history is certainly for more than 4000 years [9–11]. The highest number of mango species takes place in the Malay Peninsula, the Indonesian archipelago, Thailand, Indo-China, and the Philippines [12, 13]. Many researchers have reported that Southeast Asia is currently recognized as the center of origin due to the diversity of mango species. Thus, the origin of mango has remained controversial for many years. However, it is now grown in many tropical and subtropical regions worldwide [14, 15]. The cultivation of mango is now being distributed in about 85 countries. The important mango growing countries are Pakistan, India, China, the Philippines, Sri Lanka, Bangladesh, Indonesia, Thailand, Burma, Malaysia, Israel, Egypt, South Africa, Australia, Brazil, Cuba, and the USA [8, 16–18]. Nevertheless, the top 10 mango growing countries of the world and production scenario of 2014 are illustrated in **Figure 1**. It is shown that India is on the top followed by China, Thailand, Indonesia, Pakistan, Mexico, Brazil, Bangladesh, Nigeria, and the Philippines [19].

The mature tree of many can survive for several hundred years and attain a height of 40 meters or more [20]. The mango fruits from Pakistan are well reputed due to its delicious taste, excellent flavor, and high nutritive values [21]. Mango is providing seasonal job opportunities to most common illiterate and less literate villagers of the countryside; thus it is considered a major foreign currency earning fruit crops of Pakistan. The people from the countryside are mostly engaged in various jobs such as growing and managing the orchards, picking, packing, shipment, and processing of mango fruits. The unripe (green) mango fruits are also processed as powder form or in solar-dried slices, which is used in curies as well as in other cookies. Therefore, in the mango growing areas of Tando Qaisar, District Hyderabad, Sindh, Pakistan, it is believed to be one of the best sources of earning for local people [22].

Mango is a nutritionally rich fruit with a unique flavor, fragrance, and taste. It is an excellent source of vitamin A with flavonoids like beta-carotene, alpha-carotene, and beta-cryptoxanthin. The research reports revealed that the consumption of natural fruits rich in carotenes is known to protect against oral cavity and lung cancers. In addition, mango fruit is also a rich source of vitamins, minerals, fiber, prebiotic dietary, and antioxidant compounds, thus promoting the benefits for human health. Recent research revealed that the consumption of mango fruit protects against colon, breast leukemia, and prostate cancer [23, 24].

It is also believed that almost every part of the mango plant is used for different purposes. The mango is delighted and liked by everyone for its marvelous flavor

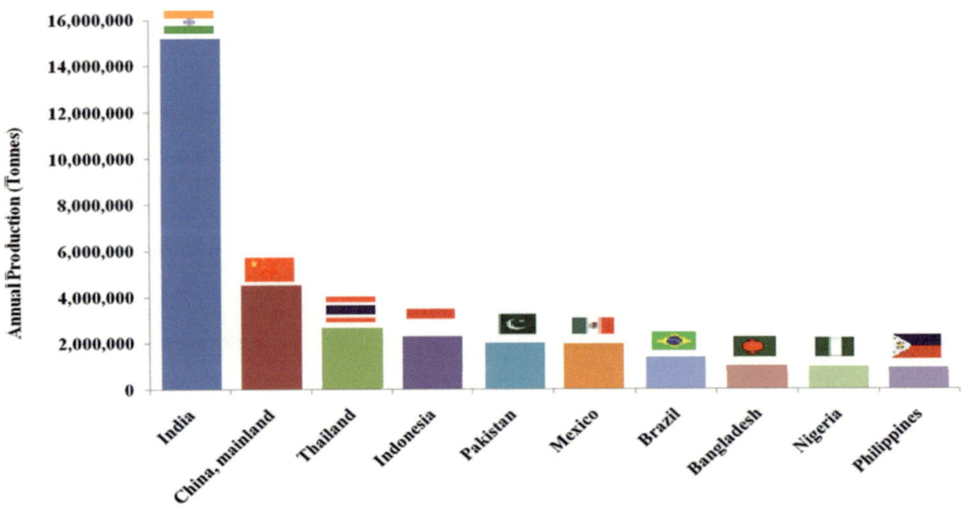

Figure 1.
Top 10 mango growing countries of the world [accessed from source: FAOSTAT database, 2014].

and delicious taste. Nutritionally as well as medicinally, mango is known to be a rich fruit. It is consumed raw as well as ripe, thus interestingly no any part is wasted. Raw fruits are sliced, dried, and floured (starch), commonly used for cooking. In addition, unripe fruit can be made into chutneys, drinks, and pickles is considered to be an effective antidote for mild forms of sunstroke, whereas ripe fruits besides being consumed as a dessert are processed into jam, squash, slices, pulp, juices, as a flavoring for baked goods, ice cream, milkshakes, jelly, marmalade, yoghurt, nectar, and mango leather. The kernel contains 8–10% fat which is used in the soap industry, while its starch can be used in the confectionary industry [6, 25].

Almost all the products and by-products made from mango are commonly used and preferred by the people in Pakistan and India. Anacardic acid is prepared from the peels (skin); although the wood quality of mango is poor, timber is used for making furniture, flooring, boats, packing cases, and other applications. Tannins mainly used for curing leather are also obtained from the bark of the tree. The twigs as well as young vegetative leaves are used for various religious purposes and for medicinal value [6, 8].

2. Major yield threats to mango production

Several pests, diseases, and disorders have been recorded on various mango varieties, ultimately resulting in severe loses to all parts of the mango around the world. Approximately 260 pest species including major and minor pests have been recorded from seedlings to mature trees at harvest and postharvest stages [26].

2.1 Mango diseases

Mango suffers from several infectious diseases caused by many phytopathogens. More than 83 different diseases and disorders including 52 fungal, 3 bacterial, and 3 plant-parasitic nematodes of the mango tree and fruit have been recorded worldwide which cause losses; however, fortunately no single disease is caused by the virus till now in mango [27, 28]. Twenty-seven diseases have been reported in mango trees of Pakistan [28]. Among them the main diseases are anthracnose, *Colletotrichum gloeosporioides* (Penz.) (**Figure 2**); powdery mildew, *Oidium mangiferae* (Bert.) (**Figure 3**); malformation, *Fusarium* spp. (**Figure 4**); bacterial leaf spot, *Erwinia mangiferae* (Doidge) (**Figure 5**); crown gall, *Agrobacterium tumefaciens* (**Figure 6**); sooty mold, *Capnodium mangiferae* (**Figure 7**); fruit rot, *C. gloeosporioides* and *Aspergillus niger* (**Figure 8**); root rot, *Rhizoctonia solani* (Kuhn) and *F. oxysporum* (Schl.) (**Figure 9**);

Figure 2.
Anthracnose disease (Colletotrichum gloeosporioides Penz.).

Figure 3.
Powdery mildew disease (Oidium mangiferae Bert.).

Figure 4.
Mango malformation disease (Fusarium spp.).

Figure 5.
Bacterial leaf spot (Erwinia mangiferae Doidge).

Figure 6.
Crown gall (Agrobacterium tumefaciens).

Figure 7.
Sooty mold (Capnodium mangiferae).

Figure 8.
Fruit rot (C. gloeosporioides and Aspergillus niger).

Figure 9.
Root rot (Rhizoctonia solani Kuhn and F. oxysporum Schl.).

dieback, *Diplodia netalensis* and *Lasiodiplodia theobromae* (Pat.) Griff. and Maubl. (Syn. *Botryodiplodia theobromae* Pat.) (**Figure 10**); gummosis, *L. theobromae* (**Figure 11**); and mango sudden decline (**Figure 12**), a complex disease [28–38].

Moreover, mango decline, mango sudden decline syndrome (MSDS), mango sudden death syndrome (MSDS), and mango tree mortality (MTM) are the common terms/phrases used for mango sudden decline disease. In the last decade, MSDS remained the most common and destructive diseases throughout Pakistan. Different workers have isolated various fungi and another organism from an infected mango tree. In a recent study, 21 different species of diseases from infected mango trees have been reported. However, it is a complex problem and actually is a result of anthracnose, dieback, root rot, tip dieback, gummosis, and dying of plants, observed very commonly in the orchards of different ages in Sindh province [22].

Figure 10.
Dieback (Lasiodiplodia theobromae Pat.)

Figure 11.
Gummosis (Lasiodiplodia theobromae).

Figure 12.
Mango sudden decline.

3. Fungicide management and resistance development

Plant disease control has always been a challenge for the growers and farmers to get optimum production due to pesticide resistance. Resistance to fungicide in current days is a major threat to plant disease management. In many plant pathogens, resistance could develop naturally; thus, several newly developed chemistries of fungicide remain at high risk. However, research toward the increase of resistance and delay in disease development has been undertaken. The existing fungicide chemistry, sometimes, renamed with new trade name does not satisfy the farmer to apply such fungicides [39] for the management of diseases. It is also obvious that even within the same fungal group the nature of resistance can be different. Fungi may develop alternative biochemical pathways around the ones that the fungicides are blocking. The blocked biochemical pathway may also be overwhelmed by the overproduction of precursors. Fungal cells may develop mechanisms to block entry of the fungicide and/or efficiently export the chemical out of cells. The result manifests itself the same way through disease control failure. An example is benzimidazole fungicides which when first introduced were more effective, and even at lower rates, than anything else on the market. The new systemic nature of the compounds and their broad spectrum of activity encouraged the wide use of these fungicides. These fungicides could control many different fungal diseases including powdery mildews, botrytis blights, leaf spots and blights, and root rots on a large variety of crops. There is even nontarget activity on other organisms including earthworms. Benzimidazole fungicides act by blocking the polymerization of tubulin preventing the nuclear division of fungal cells. The mode of action is so specific within the fungi that simple natural mutations allowed for the development of resistant fungi. The resistance is also very stable within the population. Widespread use of benzimidazole fungicides caused widespread development of resistance. Another example is dicarboximide fungicides which when first introduced were highly effective against diseases caused by *Botrytis*, *Monilinia*, *Sclerotinia*, and similar organisms. The greatest activity of these compounds is in preventing mycelial growth, but spore germination is also reduced. The resistant population is not as stable or competitive as sensitive populations and declines gradually after the selection pressure is removed. How long a break is necessary before the effective reintroduction of the dicarboximide is debatable [39].

Mango diseases as discussed earlier are sometimes hard to manage due to the pathogen cycle and perpetuation in soil and deep root. Several mango diseases are attacking from seedling to maturity; and pre-harvest to postharvest depends on the environmental conditions of the region. Chemical fungicides are believed to be a significant way to control fungal pathogens [40] or sometimes to inhibit and prevent the development and spread of pathogen [41, 42]. However, due to resistance development in pathogen or sometimes in the environment-development of new physiological races, it is hard to manage the plant diseases. The impacts of fungicide management in some importance mango diseases are discussed here.

4. Pre-harvest diseases

4.1 Anthracnose: *Colletotrichum gloeosporioides* (Penz.)

4.1.1 Cause, disease cycle, and symptoms

Anthracnose, caused by a fungal pathogen *Colletotrichum gloeosporioides*, is a severe outbreak which can cause huge economic losses at various growth stages of mango production ranging from the blossom period to postharvest. It is considered to be the most important disease of the crops in all mango producing areas worldwide [43, 44, 52]. Anthracnose is favored by high relative humidity and abundant rainfall that help in the development of the severe symptoms on leaves, flowers, fruits, and branches of all ages. The disease can cause losses varying from 50 to 100% in unmanaged orchards under a favorable environment [43, 45–46, 52]. Symptom appeared on leaves consisted of angular dark spots, about 3–5 mm long, which can coalesce, and the necrotic areas become more extended generally bordered by a yellow chlorotic halo. The leaf spots appeared due to anthracnose which serves as an important source of inoculum for the more destructive phases of infection on blossoms and fruits [47]. The conidia of *C. gloeosporioides* are disseminated by rain and/or irrigation water. Thereafter, conidia are become attached with panicles, leaves, and branch terminals at the infection site. Conidia germinate on immature fruits and tender tissues and then penetrate through the cuticle and epidermis. The fungus grows rapidly forming elongated brown necrotic lesions, which later on get blighted and ruptured. Dark blister-like spots also appear on young twigs and leaves. The leaves dry up slowly and ultimately fall down, leaving a black scar on the twig [7]. In the inflorescence, it appears as tiny black spots, which cause extensive necrosis of flowers, and small fruits fall off easily because of wind or rain leaving only the rachis attached to the tree. Affected fruits in early development can remain mummified in necrotic panicles or be aborted altogether. In the case of fruits nearing maturity, the infections are quiescent and cause irregular dark spots that quickly rot the pulp of the fruit when it reaches senescence. In the case of mature fruits, penetration occurs through the cuticle but remain quiescent until the onset of ripening, which can cause postharvest anthracnose after harvesting. In young vegetative stems, it causes canker lesions [48, 49]. Many cycles of the disease can occur as the fungus continues to multiply during the favorable seasons. The pathogen survives between seasons on infected and defoliated branch terminals and mature leaves.

4.1.2 Fungicide management

The use of fungicides for the management of anthracnose disease has been widely done worldwide. Generally, such kinds of fungicides are used when especially anthracnose is out of control due to wet and humid conditions in most commercial mango production situations [50]. To produce commercial market quality fruit, chemicals such as benomyl, copper, and mancozeb have been sprayed weekly on the flowers and at 2- to 3-week intervals on fruit until harvest [51]. Although some mango cultivars are moderately tolerant, none are sufficiently resistant to be produced without fungicides in humid areas [50]. During rainy seasons numerous preventive fungicide applications in the field are necessary to obtain acceptable fruit production [52]. In extreme situations, where fruit develops completely under disease-favorable conditions, up to 25 sprays of fungicides have been reported. It is also mentioned that fungicides are highly effective for anthracnose control reducing the severity in treated fruit by over 90% [52]. Research in India showed that fungicides reduced the severity from 54% infected fruit to 5% [53]. Research has clearly shown that low postharvest decay is associated with effective protection of fruit throughout the growing season [54]. The fungicides carbendazim, prochloraz, and benomyl tested against anthracnose on 40-year-old mango trees revealed that carbendazim showed minimum disease severity when first spraying was done during the emergence of new flushes; second and third spray applications were made at 15-day intervals. Fourth and fifth spray applications were done before monsoon (June) and after the monsoon (first week of October). Reduction in disease severity with the increased average number of fruits per tree on the third year was recorded after application of fungicides [55]. In a field trial, different doses of systemic fungicide, azoxystrobin at 1, 2, and 4 ml L^{-1} were evaluated against anthracnose disease. All doses of azoxystrobin suppressed the development of both panicle and leaf anthracnose with more production of fruits than control. The results further showed that 2 ml L^{-1} proved effective against disease than other doses of fungicides [56]. Recently in Pakistan Nasir et al. [57] evaluated seven different organic and one inorganic fungicides for their effectiveness against anthracnose disease. Three bloom sprays were applied: first at 25% flowering and two later applications at 15-day intervals. Best results were achieved with Nativo 75% WDG (tebuconazole+ trifloxystrobin) which controlled anthracnose by 92.03% and powdery mildew by 90.19%. It was followed by Cabriotop 60% WDG (metiram +pyraclostrobin) which reduced the incidence of these diseases by 89.08% and 88.04%, respectively, whereas Topsin-M 72% WP (thiophanate-methyl), Score 25% EC (difenoconazol), and Shincar 50% SC (carbendazim) provided less than 80% control. In general, all the fungicide treatments significantly reduced the incidence of the diseases and produced a higher yield of quality fruits than control in both years.

Fungicidal resistance/sensitivity among the six different isolates of *C. gloeosporioides* (Cg1 to Cg7) collected from Agricultural Export Zone (AEZ) of Andhra Pradesh and one from Tamil Nadu was studied using four systemic fungicides, viz. carbendazim (50 ppm), thiophanate-methyl (50 ppm), propiconazole (25 ppm), and hexaconazole (25 ppm), and two non-systemic fungicides, viz. mancozeb (1000 ppm) and copper oxychloride (1000 ppm), in poisoned food technique. All isolates were highly sensitive to sensitive for systemic fungicides except Cg3 which was moderately resistant to thiophanate-methyl. Isolates Cg1, Cg3, and Cg6 were highly sensitive, Cg5 and Cg7 were resistant, and Cg2 and Cg4 were highly resistant to mancozeb. It was also confirmed that all isolates were resistant to copper oxychloride. These results indicated the differential resistance/sensitivity to commonly used fungicides against *C. gloeosporioides* and allowed to recommend a specific fungicide on a regional basis [58].

However, in recent years growers have experienced problems controlling this disease, and they have suggested that the fungicides used are not providing acceptable levels of control. Products currently registered for pre-harvest use include mancozeb, copper hydroxide, and copper sulfate products—these are routinely used from flowering through to harvest. Prochloraz is used when weather conditions favor disease development, and a strobilurin product has recently been registered. Thus, an experiment was conducted to develop an integrated crop management (ICM) practice for controlling anthracnose (*C. gloeosporioides*) of mango with emphasis on nonchemical means and achieving higher yield. Pruning + weeding + spading + fertilizer + Dithane M-45 or garlic extract (three times) + irrigation (at 14-day intervals) resulted in the highest fruit retention, healthy fruits, and highest yield [58]. Postharvest hot-water treatments (15 minutes at 5°C (124–125°F) have been shown to reduce anthracnose development in ripe fruits of some specific cultivars such as "Larravi" in Puerto Rico [59] and with the cultivars "Zill," "Haden," "Sensation," "Kent," and "Keitt" for 5 minutes at about 55°C (131°F) and 15 minutes at 49°C (120°F) in Florida [60]. Hot water dips also reduced stem-end decay caused by several fungi [61]. Because of varietal differences in heat tolerance, tests must be conducted to determine the optimum time and temperature for each cultivar. However, fruit should be ripened before refrigerating in order to avoid chilling injury. Benomyl and thiabendazole at 500–1000 ppm heated to 52°C (126°F), in which mango fruits were dipped for 1–3 minutes, were effective in controlling postharvest decay on "Tommy Atkins" and "Keitt" [62–64]. However, within a short time, the fungus developed resistance to benomyl and had cross-resistance to the related fungicides thiabendazole and thiophanate-methyl (18). Heated iprodione [64], unheated prochloraz [62], and unheated imazalil [60] have also shown efficacy in controlling anthracnose. Anthracnose is best controlled by a combination of preventive measures, field fungicide sprays, and postharvest treatments.

4.2 Powdery mildew: *Oidium mangiferae* (Bert.)

4.2.1 Cause, disease cycle, and symptoms

Mango powdery mildew disease is caused by a fungal pathogen, *Oidium mangiferae* Berthet. The disease was first recorded on mango during 1914 in Brazil, and the fungus was named by Berthet [65, 66]. The associated pathogen *O. mangiferae* belongs to the class *Ascomycetes*, order *Erysiphales*, and family *Erysiphaceae* [67, 68]. This pathogen was previously considered minor; however, it became severe and attacked nearly all varieties throughout the mango growing countries of the world [69–74]. Powdery mildew of mango is favored by cool nights with warm humid weather conditions that support the severe disease incidence [30–31]. The disease has been reported in many countries of Asia, Africa, America, and Oceania [75]. Powdery mildew appears in foliage, fluorescence, petioles, young fruits, and tender stems as well. The fungus is ectophytic and reaches into the cell through haustoria by penetrating the epidermal layer. Generally, it exhibits superficial whitish fungal growth just like talcum powder. Initially, septate mycelium is developed and later on conidia, which fall after reaching maturity. On a whole, powdery mildew fungus produces white dense coating over the surface of the host. The incidence caused by this disease varied from 31.50 to 93.00% on different sensitive mango varieties. The pathogen develops in dry and cold environments but reaches greater severity at 90% relative humidity (RH) and 20–25°C [74–80].

4.2.2 Fungicide management

Several organic and inorganic fungicides have been evaluated for their effectiveness against powdery mildew disease. The powdery mildew was controlled by applying different fungicides such as benomyl, bitertanol, carbendazim, dinocap, oxythioquinone, thiophanate-methyl, tridemefon, tridemorph, Vigil, wettable sulfur, etc. [81–89]. However, some chemical fungicides such as benomyl, dinocap, and mancozeb were set up most operational at the time of flower cluster expansions and before the cluster opening [90–92]. The higher efficacy of carbendazim against powdery mildew was also proven when sprayed three times with the interval of 15 days [93]. Recently, Topas 100% EC (penconazole) and Vangard 25% EC (triadimenol) were found effective against powdery mildew (89.96% and 91.87%, respectively) [57]. In another experiment, three sprays of penconazole at prebloom, full bloom, and after fruit setting against powdery mildew, *O. mangiferae*, were applied on Samar Bahisht cultivar. It was found more effective than pyrazophos. Several other fungicides have also been tested against powdery mildew in tropical and subtropical regions of the world as protective and curative measures. In trials carried out in 1992, Topsin-M (thiophanate-methyl) was almost as effective as penconazole [94]. Trifloxystrobin has been mentioned as a new strobilurin fungicide, and it was highly effective in controlling powdery mildews on mango [95]. Some foliar fungicides, viz. Baytan Foliar, Calixin, Topas, and Bayleton, against mango powdery mildew (*O. mangiferae* Bert.) were also evaluated in 2000. Two spray applications at 15 days intervals were done on Sindhri, Siroli and Samar Bahisht, and Chaunsa varieties and revealed that Bayleton was found to be the most effective for control of disease followed by Calixin, Baytan Foliar, and Topas [79]. The study on different fungicides such as Topsin-M (thiophanate-methyl), Dithane M-45 (mancozeb), Antracol (propineb), Thiovit (sulfur), Topas (penconazole), Nordox (darosal, copper hydroxide), Anvil (hexaconazole), Aliette (fosetyl), and Rubigan (fenarimol) were tested. Maximum disease control with thiophanate-methyl and sulfur was observed. Two applications at 20–30% and 30–40% in flowering stage were used at fortnight intervals [96]. Another trial tested on Dusehri cultivar against the disease with Anvil (hexaconazole), Spotless (diniconazole), Bayleton (triadimefon), benomyl, Score (difenoconazole), and Folicur (tebuconazole) fungicides were evaluated. Anvil, Spotless, and Bayleton gave better results than other fungicides [97], whereas Score (difenoconazole) and Anpower (hexaconazole) resulted to be most effective [98]. The in vitro efficacy of hexaconazole at 0.01% and wettable sulfur at 0.3% was highly effective in the inhibition of conidial germination [99].

Six fungicides such as carbendazim (Bavistin 50 WP) 0.05%, wettable sulfur (Sulfex 80 WP) 0.25%, triadimefon (Bayletan 25 WP) 0.05%, thiophanate-methyl (Roko 70 WP) 0.1%, penconazole (Topas 10% EC) 0.05%, and hexaconazole (Contaf 5EC) 0.05% went through a 3-year experimentation (2006–2008). All the fungicides reduced the disease significantly when applied at prebloom, 10 days after first spray, and at fruit setting stage compared to the untreated control. However, hexaconazole gave the lowest incidence of powdery mildew (21.2%) and was significant over the rest of treatments except triadimefon [100]. Fungicides mostly are of high efficiency in the management of plant diseases [101]. However, the need for reducing pesticide residues in food crops, pressures to maintain a healthy environment, and often the unavailability of commercially acceptable resistant plants intensify the need for alternative methods for disease control. One of the potential methods of reducing the severity of powdery mildew in an environmentally safe manner is the use of inorganic salts like foliar spray by potassium salts [102, 103] as biocompatible fungicides. Monopotassium phosphate and potassium dihydrogen phosphate sprayed alone or in alternation with fungicides have been successful in the control of powdery mildew diseases in apples, grapes, peaches, nectarines, greenhouse cucumbers, roses, melons, and mangoes [103].

It is apparent that several studies have been conducted and workers used different fungicides successfully in order to control powdery mildew disease of mango accompanied by a high yield/tree [104]. However, fungicide resistant races of powdery mildew pathogens have been reported on several crops like as cucumber, grape vines, etc. [105–108]. The fungicide Punch was the most effective on mango and mustard when tested to managing powdery mildew disease in India [109]. However, once resistant strains appeared, most of them survived for several years, therefore the risk of re-enhancing a resistant population with further applications of ineffective [110].

4.3 Malformation: *Fusarium* spp.

4.3.1 Cause, disease cycle, and symptoms

Mango malformation disease (MMD) is a serious threat and is significantly increasing because of the great demand for mango in the international market and expansion of mango production worldwide for export [111]. It occurs in almost all mango growing countries of the world and causes severe economic losses every year [112, 113]. Much more studies are carried out on physiological, viral, fungal, acarological, and nutritional causes [114]. Mango malformation is of two distinct types, vegetative malformation (VM) and floral malformation (FM). The floral malformation is more prevalent in bearing mango trees, whereas vegetative malformation mostly appears on seedlings [115]. MMD causes shortened inflorescence, sterility, and aborted hermaphrodite flowers, and the male flowers increase in number and size [116–119]. MMD was first reported in India in 1891 [120]. It is found elsewhere in Asia (Israel, Malaysia, and Pakistan) [120], Africa (Egypt, South Africa, Sudan, Swaziland, and Uganda) [121], and Americas [Brazil, El Salvador, Mexico, Nicaragua, the USA, and Venezuela [122, 123]. In Pakistan, Khaskheli et al. [38] confirmed pathogenicity of *Fusarium nivale* (Fr.) Ces., as the first record in Pakistan and also the first report of its association with mango malformation disease in Sindh, Pakistan. In another study, Iqbal et al. [119] isolated four fungi, viz. *F. mangiferae, F. pallidoroseum, F. oxysporum*, and *Alternaria alternata*, from malformed mango parts. Akhtar et al., [117] verified the association of *F. moniliforme* and *G. fujikuroi* with the disease. Various research lines have explained the etiology and control measures of MMD; however, nature and management of this disease is still a big challenge for the researcher.

4.3.2 Fungicide management

Several attempts were made to manage the MMD; however, differences in the reported species sometimes make it difficult to control the extent of MMD. Some research lines indicate that broad-spectrum systemic fungicides are beneficial for the control of the disease [124]. Kumar et al. [125] found that mangiferin metabolites of mango induced changes in isolates of *F. moniliforme* (*Gibberella fujikuroi*). In a study on management of mango malformation through physical alteration and chemical spray, the treatment with clipping at 45 cm distance followed by a spray of benomyl results in 70.37% decrease over previous years count [113]. The best management of MMD was obtained through the treatment with clipping at 45 cm distance followed by a spray of benomyl 50 WP. Eight different fungicides (Benlate, carbendazim, Score, Daconil, captan, Topsin-M, copper oxychloride, and minimum inhibitory concentration of MICs) were applied in vitro against *F. mangiferae* of MMD. As a result, Benlate and carbendazim suppressed 100% colony growth of fungus than all other fungicides [113, 126]. In Keitt tree trunk, fosetyl-Al was injected and found to reduce the floral malformation from 96 to 48% but no effect on fruit yield [127]. The spray of Benlate and biological antagonists *Trichoderma harzianum* and *Aspergillus flavus* were also applied. The results showed that Benlate alone and in combination with antagonists gave better results in a reduction of malformation intensity [128]. Moreover, the use of insecticides, fungicides, and plant growth regulators in combination with pruning has been reported an effective integrated management measure of reducing the intensity of inoculum of MMD in the orchard [120, 128]. It is recommended that once the disease is reported in the orchard, symptomatic parts should be removed to limit the occurrence of disease [21, 130, 131]. This sanitation practice leads to a reduction in mango malformation by limiting the inoculums. However, it is difficult to impose on the large trees with panicles that are difficult to access [127, 129, 132]. In most of the mango orchards, the general practice followed to control the severity of the disease is by removal of malformed (FM and VM) parts from trees and burning them outside the orchards. Usually, pruning involves the removal and burning of infected parts. The disease occurrence could almost be reduced by following this practice at least for 2–3 consecutive years. More recently, in vitro and in vivo attempts were made with commercial fungicides to reduce the severity of *Fusarium nivale* (Fr.) Ces., a predominant and virulent fungus in mango orchards of Sindh, Pakistan, which was first isolated from mango MMD [38, 133]. Mycelial growth of *F. nivale* was significantly inhibited at low doses of thiophanate-methyl and fosetyl-Aluminum. Metalaxyl+mancozeb and mancozeb alone also reduced the growth of fungus at their high doses, respectively, as compared to copper oxychloride. Thiophanate-methyl and fosetyl-Aluminum significantly reduced infection in Desi (local), Almas, and Dusheri after the first spray. The second spray of thiophanate-methyl and fosetyl-Aluminum fungicides completely inhibited infection of *F. nivale* and 100.0% reduction in vegetative malformation disease in Desi, Almas, and Dusheri, as compared to metalaxyl+mancozeb, copper oxychloride, and in control. The application of thiophanate-methyl and fosetyl-Aluminum would be useful in integrated management of MMD [133].

4.4 Sooty mold: *Capnodium mangiferae* (Cooke and Broome)

4.4.1 Cause, disease cycle, and symptoms

The sooty mold is a fungal disease caused by *Capnodium mangiferae* (Cooke & Broome) [134]. In actual, the sooty mold-causing fungi establish its growth on sugary secretion by insects. Thus, occurrence, incidence, and severity of this disease depend on the infestation of insect pests. Usually, sucking insect pests infest in the beginning of disease occurrence and excrete sweet secretion upon which sooty mold develops. Later on, sooty mold appears as a black velvety growth on the leaf surface. The entire leaf surface or portion of the leaf may be covered with fungal growth, and in severe cases, the whole plants are affected. The thin layer formed on the leaf surface can be rubbed off easily [135]. When the molds attack blossoms during flowering time, the fruit set is affected, and in severe cases, small fruits also fall down [114]. Under the dry conditions, this may be blown off as small fragments by the wind. The disease-causing fungi in true sense are nonpathogenic; however, photosynthetic activity of the plant is impaired due to the covering of the leaves. The symptoms occurring due to this disease are quite obvious in diseased orchards [135].

4.4.2 Fungicide management

It has already been explained that sooty develops on the exudates of sucking insect pests such as aphids, leafhoppers, mealybugs, psyllids (including eucalyptus lerp psyllid), soft scales, and whiteflies. Both the immature and adult stages of these insects feed by sucking sap from plants, producing honeydew. Their common characteristic is that they all suck sap from plants. Therefore, these pests need serious attention and should be controlled through pesticides.

It is also pertinent to mention that sooty mold-causing fungus is a weak parasite; therefore, sometimes uses of fungicides are necessary to apply. The effectiveness of seven organic fungicides, one synthetic fungicide, bagging of fruits, and the untreated control were evaluated to control sooty mold on leaves and fruit of mango "Manila," in Veracruz, Mexico. Results showed that the bio-fungicides Bio hcaz 3.5, Bio fyb 1.5, Fungicus ph 4 y Fungicus ph 8 provided 95% of leaves in the categories of healthy and light (less than 5% damage). Percentage of healthy fruits was 98% for bagging, 82% for benomyl, 80% for Sunset 3, and 78% for Sulfocop 4 and Bio fyb 1.5. Bio fyb 1.5 showed good control of sooty mold in leaves and fruits. The application of these organic products did not have a negative effect on the yield and fruits quality [136].

4.5 Mango sudden decline

4.5.1 Cause, disease cycle, and symptoms

Mango decline or mango sudden decline syndrome or mango sudden death syndrome or mango tree mortality are some common terms/phrases used for this disease. This remained the most common and destructive disease throughout Pakistan in the recent past years. In actual, MSDS is a complex disease of mango, and several pathogenic fungi were isolated from this complex problem [29–38]. Different investigators have isolated various fungi and another organism from an infected mango tree. Ahmed et al. [137] reported that the onset of dieback become evident by discoloration and darkening of twigs from the tip downward due to *Diplodia natalensis*. Ploetz et al. [138] observed the symptoms of decline, tip dieback, and gummosis from mango nurseries artificially inoculated with *A. alternata*, *G. cingulata*, *D. dominicana*, *L. theobromae*, and *Phomopsis* sp. According to Saleem and Akhtar [32], it is caused by root rot, anthracnose, and dieback. Jiskani [22] reported MSDS as disease complex caused by the combined attack of several different fungi and abiotic factors. Shahbaz et al. [139] pointed out disorders like twig blight, gummosis, bark splitting/cracking, and wilting as a cause. Leghari [140] isolated 10 genera with 12 species of different fungi from infected mango trees showing sudden death syndrome symptoms; however, *L. theobromae* (Prv. *B. theobromae*) was predominant with 20–83.30% incidence and 62.50–85% severity followed by *F. solani* with 56.66–73.33% incidence and 62.50–78.75% severity of mango decline. Hakro [141] isolated nine plant-parasitic nematodes and seven fungi from roots of the dead mango tree and studied the interaction of most predominant nematode with fungi (*Xiphinema index*, *F. oxysporum*, and *R. solani*). It is reported that gum is the most common symptom and *L. theobromae* was the most abundant isolated fungus, whereas it is also reported to be caused by *Ceratocystis fimbriata*. Asad et al. [144] isolated *L. theobromae*, *C. fimbriata*, and *Phomopsis* sp. as the most common fungi causing MSDS. Similar associations between *L. theobromae* and mango decline have been observed by quite a number of researchers [145–151]. Recently, the incidence of MSD was found to be 20% in Punjab and more than 60% in Sindh Provinces of Pakistan and 60 percent in Al Batinah region of Oman [152]. This phenomenon has also been reported from some other parts of the world, i.e., Brazil and Oman [152, 153]. In Brazil, Oman, and Pakistan, *C. fimbriata*, *C. omanensis*, and *L. theobromae* were the main causal organisms of MSDS were mostly isolated from diseased mango tree [152–156]. It has also been reported that fungus is mainly disseminated by the mango bark beetle, *Hypocryphalus mangiferae* (Stebbing), by infected plant material and the infested soils where it is able to survive for long periods. The best way to avoid losses due to MSD is to prevent its establishment in mango production areas [157]. *Ceratocystis manginecans* is also reported as the causal agent of a destructive mango wilt disease in Oman and Pakistan [158]. However, according to Jiskani et al., [22] it is a complex problem and actually is a result of anthracnose, dieback, root rot, tip dieback, gummosis, and dying of plants, observed very common in the orchards of different ages in Sindh. Moreover, studies also revealed that some mango varieties had more prevalence of MSDS than others, and the age-wise differences were also noticed. The findings indicate that trees of some varieties have the ability to overcome MSDS at a specific age and the severity of MSDS may be manageable till that. The variable incidence in different mango varieties at different ages can also be attributed to a possible natural tolerance against MSDS [159].

4.5.2 Fungicide management

There have been several attempts to manage this severe outbreak of mango; however, the recovery of the affected tree sometimes could not succeed due to the development of fungi in the vascular system of tree. All reported fungi of MSD are soilborne and cause tree decline or wilt after several years of development. Some studies showed that Topsin-M and Daconil are the most effective fungicides, whereas copper oxychloride intermediate and mancozeb were the least effective in inhibiting the mycelial growth of *B. theobromae* under in vitro conditions [160]. In another report, Topsin-M and Aliette when applied to infected tree trunks through the injection method, three times at an interval of 10 days, are reported as the most effective fungicides.

4.6 Dieback: *Lasiodiplodia theobromae* (Pat.)

4.6.1 Cause, disease cycle, and symptoms

Mango dieback caused by *L. theobromae* is considered one of the serious diseases of mango. Previous research has already proven *L. theobromae* as the cause of dieback of mango [161–162]. *L. theobromae* is a soilborne fungus causing both field and postharvest diseases in about 2080 plant species [161, 163–166]. In Pakistan, it has been reported on more than 50 species of plants [167]. Among several diseases responsible for low crop production in Pakistan, mango dieback is also one of them. To further complicate the issue, resistance in *L. theobromae* is emerging against different fungicides [168]. Thus, this disease has been described as one of the serious threats to mango growers [154]. The onset of dieback becomes clear by discoloration and darkening of twigs, oozing of gum, wilting of leaves, dieback, browning of vascular bundles, and death of the entire plant [147, 162, 169]. Symptoms can also be observed in reproductive structures [170]. In severe situations, branches start drying one after another in a sequence resulting in the death of the trees of the mango plantation. Commonly, once the symptoms of decline or widespread dieback are evident, it is difficult to stop or reverse the progress of the disease. The disease has also been observed on different mango varieties associated with the variation in their susceptibility toward the fungus. Reports have shown that certain varieties are highly susceptible [171–172]. In vivo studies demonstrated that *L. theobromae* becomes aggressive in colonizing host tissues when plants are under abiotic stress, such as heat, water stress, or drought stresses [173–174]. In general, dieback is one of the deadly diseases of mango, which causes serious damage to the tree and its productivity [175].

4.6.2 Fungicide management

Several attempts are made to control the mango dieback throughout the mango growing countries of the world. In Pakistan, mycelial growth of *L. theobromae* was significantly inhibited by carbendazim and thiophanate-methyl when used at 1 ppm a.i. or more. Alliete was effective at relatively high concentrations, i.e., at 1000 and 10,000 ppm a.i., whereas Copxykil, Cuprocaffaro, and Thiovit failed to inhibit the mycelial growth of *L. theobromae*. In the field experiment, carbendazim was found to be more effective than thiophanate-methyl and Alliete in reducing the fungal infection in mango plants, suppressing the gum exudation, dieback, and wilting, resulting in a significant enhancement in vegetative growth of plants [142, 143]. The efficacy of different fungicides, viz. copper oxychloride, diethofencarb, pyraclostrobin, carbendazim, difenoconazole, mancozeb, and thiophanate-methyl, was evaluated in vitro using a poison food technique. Thiophanate-methyl at all concentrations was found to be the most effective among five systemic fungicides against *L. theobromae*, followed by carbendazim, difenoconazole, and diethofencarb. The fungicides, i.e., thiophanate-methyl, difenoconazole, carbendazim, and diethofencarb, showed maximum efficacy with increasing concentration. The isolates of *L. theobromae* showed some resistance development against the tested fungicides when compared with previous work. These investigations provide new information about chemical selection for the control of holistic disease in mango growing zones of Pakistan [168]. The fungicide treatments such as cobox (copper oxychloride), precure combi (thiophanate-methyl + diethofenocarb), and Cabrio top (pyraclostrobin + metiram) at four concentrations 25, 50, 75, and 100 ppm were also evaluated in vitro against *L. theobromae*. Cobox (copper oxychloride) and precure combi (thiophanate-methyl + diethofenocarb) showed a significant reduction in colony diameter of *L. theobromae* at all concentrations compared to non-treated/control. Cabrio top (pyraclostrobin + metiram) did not significantly decrease the colony diameter of *L. theobromae.* In the field precure combi (thiophanate-methyl + diethofenocarb) and cobox (copper oxychloride) were also used as foliar fungicide and in soil drenching in addition to soil amendment with silt, NPK, and manure and pruning to manage the naturally affected mango plants with dieback, and cobox was found best where soil amendment and pruning was done with soil drenching [176]. Recently, in the UAE, some systemic chemical fungicides, Score, Cidely Top, and Penthiopyrad, significantly inhibited the mycelial growth of *L. theobromae* both in vitro and in the greenhouse. Cidely Top proved to be a highly effective fungicide against *L. theobromae* dieback disease also under field conditions [175]. Despite efforts made to control the dieback disease using different fungicides, disease incidence is still high in some farms in Oman. Testing sensitivity of 28 randomly selected *L. theobromae* isolates to four commonly used fungicides showed that the EC50 levels of the isolates were in the range of 0.01–8.75 (avg. 0.54 mg L^{-1}) for iprodione, 0.1–242.8 (111.6 mg L^{-1}) for copper oxychloride, 40.3–738.1 (avg. 229.3 mg $^{L-1}$) for copper hydroxide, and 0.1 to over 1000 (avg. > 1000 mg L^{-1}) for thiophanate-methyl. The study showed the development of resistance to some fungicides, especially with thiophanate-methyl. The development of resistance to fungicides could be one of the main reasons behind the reduction in the efficacy of managing dieback in the studied farms [177].

Author details

Muhammad Ibrahim Khaskheli
Department of Plant Protection, Sindh Agriculture University, Tandojam, Sindh, Pakistan

*Address all correspondence to: mikhaskheli@sau.edu.pk

References

[1] Litz RE, Gomez-Lim MA. *Mangifera indica* Mango. In: Litz RE, editor. Biotechnology of Fruit and Nut Crops. Wallingford, UK: CAB International; 2005. pp. 40-61

[2] Purseglove JW. Mangoes west of India. Acta Horticulturae. 1972;(24):107-174

[3] Popenoe W. The mango. In: Manual of Tropical and Subtropical Fruits. New York: The Macmillan Co; 1920. p. 135

[4] de Candolle A. Origin of Cultivated Plants. 2nd ed. London: Kegan Paul; 1904

[5] Vavilov NI. The origin, variation, immunity and breeding of cultivated plants. Chronica Botanica. 1926;**13**(6):1949-1950

[6] Chakrabarti DK. Mango Malformation. Vol. 1. NY, USA: Springer; 2011. p. 148

[7] Litz RE. The Mango: Botany, Production and Uses. Wallingford, UK: CABI; 2009

[8] Mukherjee SK, Litz RE. Introduction: Botany and importance. In: Litz RE, editor. The Mango: Botany, Production and Uses. Wallingford, UK: CAB International; 2009. pp. 01-14

[9] De Candolle AP. Origin of Cultivated Plants. Hafner, London; 1884

[10] Mukherjee SK. Origin, distribution and phylogenetic affinities of the species of *Mangifera* L. Journal of the Linnean Society: Botany. 1953;**55**:65-83

[11] Campbell RJ. A Guide to Mangoes in Florida. Miami, Florida: Fairchild Tropical Garden; 1992

[12] Mukherjee SK. Systematic and Ecogeographic Studies of Crop Gene Pools. 1. *Mangifera* L. Rome: International Board for Plant Genetic Resources; 1985

[13] Bompard JM. Wild Mangifera Species in Kalimantan (Indonesia) and in Malaysia. Final Report. Rome: International Board for Plant Genetic Resources; 1989

[14] Mukherjee SK. Introduction: Botany and importance. In: Litz RE, editor. The Mango: Botany, Production and Uses. Wallingford, UK: CAB International; 1997. pp. 1-19

[15] Deependra Y, Mango SSP. History origin and distribution. Journal of Pharmacognosy and Phytochemistry. 2017;**6**(6):1257-1262

[16] Srivastava RP. Mango Cultivation. Lucknow, Uttar Pradesh (India): International Book Distributing Company; 1998. p. 633

[17] Singh RN. The Mango-Botany. Cultivation and Utilization. London. World Crop Book Series: Leonard Hill (Books) Ltd; 1960. p. 438

[18] Knight RJ Jr, Schnell RA. Mango (*Mangifera indica* L.) introduction and evaluation in Florida and its impact on the world industry. Acta Horticulturae. 1993;(341):125-135

[19] FAOSTAT. Food and Agriculture Organization of the United Nation. Production of Mangoes, Mangosteens, and Guavas in 2016, Crops/Regions/ World List/Production Quantity (Pick Lists). Food and Agriculture Organization of the United Nation, Corporate Statistical Database (FAOSTAT); 2018. [Accessed on: 2018-08-12]

[20] Kostermans AJGH, Bompard JM. The Mangoes: Botany, Nomenclature, Horticulture,

Cultivation and Utilization. London: Academic Press; 1993

[21] Muhammad FI, Pervaiz MA. Effect of fungicides on mango malformation. Pakistan Journal of Biological Sciences. 1999;**2**:772-773

[22] Jiskani MM, Pathan MA, Wagan KH, Khaskheli MI. Documentation of identified and unidentified diseases of mango in Sindh, Pakistan. In: International Symposium on Prospects of Horticultural Industry in Pakistan. Pakistan: Institute of Horticultural Science, University of Faisalabad; 2007. pp. 176-190

[23] Hamdard MS, Rafique MR, Farroq U. Physico-chemical characteristics of various mangos, *Mangifera indica* L varieties. Journal of Agricultural Research. 2004;**42**(2):191-199

[24] Anonymous, 2018. Mango Fruit Nutrition Facts. Available from: https://www.nutrition-and-you.com/mango-fruit.html [Accessed on: 2018-08-12]

[25] Jiskani MM. Dying of mango orchards needs special attention. Pakistan and Gulf Economist, Karachi, 2002, Mar 18-24. Available from: www.pakistaneconomist.com/page/issue11/i&e3.htm. [Accessed on: 2018-03-04]

[26] Veeresh GK. Pest problems in mango-world situation. Acta Horticulturae. 1989;(231):551-565

[27] Pernezny K, Simone GW. Diseases of Mango (*Mangifera indica* L.). In: Common Names of Plant Diseases. St. Paul, MN: American Phytopatholoical Society; 2000

[28] Ghafoor AG, Khan SAJ. List of Diseases of Economic Plants in Pakistan. Karachi: Deptt. Pl. Prot; 1976. p. 85

[29] Javaid A. Major diseases of mango in Pakistan and their management- a review. Pakistan Journal of Phytopathology. 2005;**17**:108-112

[30] Kamal M, Moghal SM. Studies on Plant Diseases of South West Pakistan. Tandojam, Pakistan: Agricultural Research Institute; 1968. p. 207

[31] Hafiz A. Plant Diseases. Islamabad, Pakistan: Pakistan Agricultural Research Council; 1986. p. 552

[32] Saleem A, Akhtar KM. Mango diseases and their control. In: Choudhry AG, Syed MA, editors. Proc. Intl. Mango Workshop Feb. 27 to Mar. 01. Multan: Agri. Ext.; 1989. pp. 143-154

[33] Arshad J, Bajwa R, Anjum T. Tree dieback in Punjab, Pakistan. Mycopathologia. 2004;**2**(1):1-5

[34] Khanzada MA, Lodhi AM, Saleem S. Mango decline in Sindh: Etiology and control. Souv. 5th Meeting Intl. Symp. On Recent Trends in Plant Disease Management 20-22 Dec., 2005. pp. 78

[35] Nizamani ZA. Some studies on biology and management of tip dieback disease of mango, *Mangifera indica* L. [M.Sc. Thesis]. Tando jam: Sindli Agriculture University; 2004

[36] Pathan MA, Leghari TN, Jiskani MM, Wagan KH. Evaluation of different fungicides against Botryodiplodia theobromae causing mango gummosis. In: Souvenir on 5th Meeting of Pak. Phytopath. Soc. International Symposium on Recent Trends in Plant Disease Management. 2005. p. 68

[37] Fateh FS, Kazmi MR, Ahmad I, Ashraf M. *Ceratocystis fimbriata* isolated from vascular bundles of declining Mango trees in Sindh, Pakistan. Pakistan Journal of Botany. 2006;**38**(2):1257-1259

[38] Khaskheli MI, Pathan MA, Jiskani MM, Wagan KH, Soomro MH,

Poussio GB. First record of *Fusarium nivale* (Fr.) Ces. Associated with mango malformation disease (MMD) in Pakistan. Pakistan Journal of Botany. 2008;**40**(6):2641-2644

[39] Pscheidt JW, Ocamb CM. Pacific Northwest Plant Disease Management Handbook [online]. Corvallis, OR: Oregon State University; 2018

[40] DaSilva PAV, Martins RB, Michereff SJ, DaSilva MB, Camara MPS. Sensitivity of *Lasiodiplodia theobromae* from Brazilian papaya orchards to MBC and DMI fungicides. European Journal of Plant Pathology. 2012;**13**:489-498

[41] Malik MT, Haq CA, Grewal AG, Khan MI, Nasir M. Towards the control of a new threat "quick decline of mango". Pakistan Horticulture. 2003;**3**:37-40

[42] Karaoglanidis GS, Karadimos DA, Ioannidis PM, Ioannidis PI. Sensitivity of *Cercospora beticola* populations to fentin-acetate, benomyl and flutriatol in Greece. Crop Proctection. 2003;**22**:735-740

[43] Chowdhury MNA, Rahim MA. Integrated crop management to control anthracnose (*Colletotrichum gloeosporioides*) of mango. Journal of Agriculture and Rural Development. 2009;**1-2**:115-120

[44] Mora AA, Téliz OD. Enfermedades del mango. In. Mora AA, Téliz OD, Rebouças SJA, Editors. Mango: Manejo y Comercialización. Colegio de Postgraduados en Ciencias Agrícolas. México; 2002. pp. 55-171

[45] Benítez FA, Huerta PG, Holguín FM, Toledo JA. Efectode *Colletotrichum gloeosporioides* (Penz.) Penz. and Sacc. en lacaída de frutos de mango cv. Ataulfo en el Soconusco, Chiapas, México. Revista Mexicana de Fitopatología. 2003;**21**:223-227

[46] Akem CN. Mango anthracnose disease: Present status and future research priorities. Plant Pathology Journal. 2006;**5**:266-273

[47] Kwee LT, Chong KK. Diseases and Disorders of Mango in Malaysia. Kuala Lumpur, Malaysia: Art Printing Works SDN. BHD; 1994. p. 101

[48] Holguín MF, Huerta PG, Benítez CF, Toledo AJ. Epidemiología de la antracnosis *Colletotrichum gloeosporioides* (Penz.) Penz. And Sacc. en mango (*Mangifera indica* L.) cv. Ataulfo en el Soconusco, Chiapas, México. Revista Mexicana de Fitopatología. 2009;**27**:93-105

[49] Jeffries P, Dodd JC, Jeger MJ, Plumbley RA. The biology and control of Colletotrichum species on tropical fruit crops. Plant Pathology. 1990;**39**:343-366

[50] Ploetz RC. The major diseases of mango: Strategies and potential for sustainable management. Acta Horticulturae. 2004;(645):137-150

[51] McMillan RT Jr. Control of mango anthracnose with foliar sprays. Proceedings of the Florida State Horticultural Society. 1984;**97**: 344-345

[52] Arauz LF. Mango anthracnose: Economic impact and current options for integrated management. Plant Disease. 2000;**84**(6):600

[53] Ray SK, Das S, Hasan MA, Jha S. Efficacy of different fungicides for the management of anthracnose of mango on old and rejuvenated orchard in West Bengal. Acta Horticulturae. 2009;**820**:529-533

[54] Swart SH, Swart G, Labuschagne C. The effect of strategically timed pre-harvest fungicide applications on post-harvest decay of mango. In: Proceedings of the

VIII International Mango Symposium. Acta Horticulturae. 2009. p. 820

[55] Ray SK et al. Efficacy of different fungicides for the management of anthracnose of mango on old and rejuvenated orchard in West Bengal. In: Proceedings of the VIII International Mango Symposium. Acta Horticulturae. 2009. p. 820

[56] Sundravadana S, Alice D, Kuttalam S, Samiyappan R. Efficacy of azoxystrobin on *Colletotrichum gloeosporioides* Penz. Growth and on controlling mango anthracnose. Journal of agricultural and biological. Science. 2007;**2**(3):10-15

[57] Nasir M, Babar I, Muhammad I, Muhammad S, Muhammad Z, Niaz HA, et al. Efficacy of some organic fungicides against anthracnose and powdery mildew of Mango. Pakistan Journal of Agricultural Sciences. 2017;**54**(3):493-496. DOI: 10.21162/PAKJAS/17.1909

[58] Kumar AS, Reddy NPE, Reddy KH, Devi MC. Evaluation of fungicidal resistance among *Colletotrichum gloeosporioides* isolates causing mango anthracnose in agri export zone of Andhra Pradesh, India. Plant Pathology Bulletin. 2007;**16**:157-160

[59] Pennock W, Maldonaldo G. Hot-water treatment of mango fruits to reduce anthracnose decay. The Journal of Agriculture of the University of Puerto Rico. 1962;**46**:272-283

[60] Smoot JJ, Segall RH. Hot water as a postharvest control of mango anthracnose. Plant Disease Report. 1963;**47**:739-742

[61] Spalding DH, Reeder WF. Decay and acceptability of mangos treated with combinations of hot water, imazalil, and gamma radiation. Plant Disease. 1986;**70**:1149-1151

[62] Johnson GI, Sangchote S, Cooke AW. Control of stem end rot (Dothiorella dominicana) and other postharvest diseases of mangos (cv. Kensington pride) during short-and long-term storage. Tropical Agriculture (Trinidad). 1989;**67**:183-187

[63] Spalding DH, Reeder WF. Postharvest disorders of mangos as affected by fungicides and heat treatments. Plant Disease Report. 1972;**56**:751-753

[64] Spalding DH, Reeder WF. Controlling market diseases of mangos with heated benomy! Proceedings of the Florida State Horticultural Society. 1978;**91**:186-187

[65] Briton Jones HR. Mycological Work in Egypt during the Period 1920-22. Technical and Scientific Service Bull No. 49. Ministry of Agriculture Egypt; 1923. pp. 1-129

[66] Uppal BN, Patel MK, Kamat MN. Powdery mildew of mango. Journal University of Bombay. 1941;**9**:12-16

[67] Braun U, Cook RTA, Inman AJ, Shin HD. The taxonomy of powdery mildew fungi. In: Belanger RR, Bushnell WR, Dik AJ, Carver TLW, editors. The Powdery Mildews. St. Paul, Minnesota, USA: APS Press; 2002. pp. 13-55

[68] Agrios GN. Plant Pathology. USA: Elsevier Academic Press; 2005. p. 922

[69] Prakash O, Sirivastav KC. Mango Diseases and its Management. A World Review. Karal Bagh, New Dehli: Today and Tomorrow Printers and Publishers; 1987. p. 175

[70] Prakash O, Misra AK. Fungal diseases of sub-tropical horticultural fruit crops. In: Chadha KL, editor. Advance in Horticulture. New Delhi,

India: Malhotrapub. House; 1993. pp. 1275-1372

[71] Prakash O, Raoof MA. Studies on powdery mildew (*Oidium mangiferae*) disease of mango: Distribution, perpetuation, losses and chemical control. Biological Memories. 1994;**207**:31-45

[72] Prakash O. Post harvest disease problems of Indian mangoes and further strategies for research. Biological Memoirs. 1996;**22**:27-38

[73] Ploetz RC, Prakash O. Foliar, floral and soil borne diseases of mango. In: Litz RE, editor. The Mango. Wallingford, UK: CABI; 1997. pp. 281-325

[74] Khaskheli MI, Jiskani MM, Pathan MA,Wagan KH, Goswami SP. Occurrence of powdery mildew of mango on different varieties under agro-ecological conditions of Hyderabad. In: Proc. 3rd National Conf. Agri. & Animal Sci. Tandojam: SAU; 2007. pp. 62-65

[75] Nasir M, Mughal SM, Mukhtar T, Awan MZ. Powdery mildew of mango: A review of ecology, biology, epidemiology and management. Crop Protection. 2014;**64**:19-26

[76] Pathak VN, Khatri NK, Pathak M. Fundamentals of Plant Pathology. India: Agrobios Publisher; 2006. p. 322

[77] Mishra A, Bohra A, Mishra A. Plant Pathology Diseases and Management. India: Agrobios Publisher; 2005. p. 766

[78] Schoeman MH, Manicom BQ. Epidemiology of powdery mildew on mango blossoms. Plant Disease. 1995;**79**:524-529

[79] Jiskani MM, Wagan KH, Pathan MA, Arain MH. Efficacy of different fungicides on powdery mildew of mango caused by *Oidium mangiferae* (Berthet). Pakistan journal of agriculture, agricultural engineering and veterinary. Science. 2000;**16**(1-2):32-35

[80] Pérez-Rodríguez A, Monteón-OjedaA,AntonioMora-AguileraJ, Hernández-Castro E. Epidemiology and strategies for chemical management of powdery mildew in mango. Pesquisa Agropecuária Brasileira. 2017;**52**(9):715-723. DOI: 10.1590/S0100-204X2017000900003

[81] Lingaraj DS. Mango diseases and pests (*Mangifera indica* L.). Journal of the Mysore Horticultural Society. 1969;**14**:15-18

[82] Prakash O. Chemical control of phoma blight in mango. Indian Journal of Mycology and Plant Pathology. 1979;**9**(2):184-185

[83] Gupta PC, Dang JK. Occurrence and control of powdery mildew of mango in Haryana. Indian Phytopathology. 1981;**33**(4):631-632

[84] Prakash O, Singh UN. Evaluation of various fungicides for the control of powdery mildew of mango caused by *Oidium mangiferae*. Pesticides. 1982;**16**(2):17-18

[85] Joshi HU, Chauhan HL. Effective control of powdery mildew of mango. In: (Abs.) Sec. Int. Sym. Mango, Bangalore, India. 1985. p. 63

[86] Datar VV. Management of powdery mildew of mango with fungicides. Indian Phytopathology. 1986;**39**:271-272

[87] Rawal RD, Ullasa BA. Control of powdery mildew (Oidium mangiferae berth.) of mango by fungicides. In: ISHS Acta Horticulturae 231: II Inter. Symp. Mang. Bangalore India. 1989

[88] Tahir HAS, Sahi ST, Iqbal M, Sahi GM, Atif M. Evaluation of new fungicides against powdery mildew

of mango (*Mangifera indica*). Pakistan Journal of Phytopathology. 2009;**21**(2):126-129

[89] Ravikumar MR, Vithal N, Yashaswini S, Tippesh C. Bio- efficacy and Phyto-Toxicity of Azoxystrobin 23% SC against powdery mildew (*Oidium mangiferae*) and anthracnose (*Colletotrichum gloeosporioides*) diseases in Mango. International Journal of Current Microbiology and Applied Sciences. 2017;**6**(10):314-321. DOI: 10.20546/ijcmas.2017.610.038

[90] Ruehle GD, Ledin RB. Mango growing in the Florida Agricultural Experimental Station, Bulletin No. 1956; p. 57

[91] McMillan RT Jr. Control of anthracnose and powdery mildew of mango with systemic and nonsystemic fungicides. Tropical Agriculture. 1973;**50**:245-248

[92] Persley DM, Pegg KG, Syme JR, editors. Fruit and Nut Crops- a Disease Management Guide. Queensland Department of Primary Industries Information Series1989. p. 188018

[93] Prakash O, Raoof MA. Spray schedule for the control of powdery mildew disease of mango. Ind. Journal of Plant Pathology. 1985;**3**:51-52

[94] Haq CA, Malik MT, Syed SA, Khan SH. Evaluation of various fungicides against powdery mildew (*Oidium mangiferae*) in mango. Pakistan Journal of Phytopathology. 1994;**6**(1):17-20

[95] Moshe R. Efficacy of trifloxystrobin (Flint), a new strobilurin fungicide, in controlling powdery mildews on apple, mango and nectarine, and rust on prune trees. Crop Protection. 2000;**19**(5):335-341

[96] Akhtar KP, Khan IA, Shakir AS, Khan SM. Evaluation of fungicides against powdery mildew disease of mango. Pakistan Journal of Phytopathology. 1998;**10**(1):26-29

[97] Iqbal M, Ihsan J. Chemical control of powdery mildew of mango. Pakistan Journal of Phytopathology. 2001;**13**(1):31-33

[98] Raheel M, Anwar SA, Javed N, Ilays MB, Zia A. Management of powdery mildew of mango by foliar spray fungicides. Pakistan Journal of Phytopathology. 2008;**20**(1):173-174

[99] Mehta ND, Patel PR, Pandya HV, Patel SD. Assessment of various fungicides and bio-agents against the powdery mildew of mango (*In vitro*). International Journal of Chemical Studies. 2018;**6**(1):1063-1065

[100] Chavan RA, Deshmukh VD, Tawade SV, Deshmukh JD. Efficacy of fungicides for managing powdery mildew of mango. International Journal of Plant Protection. 2009;**2**(1):71-72

[101] Dekker J. Build- up and persistence of fungicide resistance. In: Brend EJ, Atkin EK, editors. National Pesticide use. Cambridge. U.K: Cambridge University Press; 1987. pp. 153-168

[102] Menzies J, Bowen P, Ehret D, Glass ADM. Foliar applications of potassium silicate reduce severity of powdery mildew on cucumber, muskmelon. And zucchini squash. Journal of the American Society for Horticultural Science. 1992;**117**:902-905

[103] Mahrous HAH. Effectiveness of foliar spraying with phosphates salts fertilizer, sodium bicarbonate and a fungicide in controlling powdery mildew disease of grape vine. Journal of Agricultural Science - Mansoura University. 2003;**28**(9):7043-7049

[104] Reuvehi M, Harpez RM, Reuveni R. Integrated control of powdery mildew on field-grown mango

trees by foliar sprays of mono-potassium phosphate, sterol inhibitor fungicides and strobilurin kresoxym-methyl. European Journal of Plant Pathology. 1998;**104**:853-860

[105] Schroeder WT, Prowidenti R. Resistance in powdery mildew of cucurbits. Plant Disease Report. 1969;**53**:27

[106] Schepers HTAA. Decreased sensitivity of Funliginea to fungicides which inhibit ergosterol biosynthesis. Plant Pathology. 1983;**89**:185-187

[107] Steva H, Cherjeau M. Cross resistance biosynthesis inhibitor. Meded Fac Landbouwwet Rijksuniv Gent. 1990;**55**:983-988

[108] McGrath MT. Increased resistance to triadimefon and to benomyl in *Sphaerotheca fuliginea* populations following fungicide usage over one season. Plant Disease. 1996;**80**:633-639

[109] Kaushik CD, Mishra BP. Flusilazole (punch), a potential fungicide for the control of different powdery mildews. Annals of Agri Bio Research. 1997;**2**(1):15-19

[110] Azza M, Azmy K. Controlling of mango powdery mildew by some salts, growth regulators and the biofungicide AQ10 compared with punch fungicide in Egypt. American Journal of Life Sciences. Special Issue: Role of Combination Between Bioagents and Solarization on Management of Crown- and Stem-Rot of Egyptian Clover. 2014;**2**(6-2):33-38. DOI: 10.11648/j.ajls.s.2014020602.15

[111] Ahmad F, Hafiz IA, Isi AA, Ahmed S, Khan M. Mango varietal susceptibility to malformation and its control. Asian Journal of Plant Sciences. 2002;**1**(2):158-159

[112] Iqbal Z, Saleem A, Dasti AA. Assessment of mango malformation in eight districts of Punjab (Pakistan). International Journal of Agriculture and Biology. 2004;**6**(4):620-623

[113] Iqbal Z, Akhtar N, Ghazanfar MU, Shehzad SM, Ahmad S, Asif M, et al. Management of mango malformation through physical alteration and chemical spray. African Journal of Agricultural Research. 2011;**6**(7):1897-1901

[114] Prakash OM. Compendium of Mango Diseases and Disorders. New Delhi, India: Capital Publishing Company; 2003. p. 84

[115] KhaskheliMI,PathanMA,JiskaniMM, Wagan KH,Goswami SP. Survey, intensity and varietal response of mango malformation disease (MMD) in different Mango orchards of Sindh, Pakistan. Pakistan Journal of Agriculture, Agricultural Engineering and Veterinary Sciences. 2008;**24**(1):53-56

[116] Kumar J, Singh US, Beniwal SP. Mango malformation: One hundred years of research. Annual Review of Phytopathology. 1993;**31**:217-232

[117] Akhtar KP, Khan IA, Jaskani MJ, Asif M, Khan MA. Isolation and characterization of *Fusarium moniliforme* var. *subglutinans* and *Gibberella fujikuroi* var. *subglutinans* from malformed mango. Sultan Qaboos Uni. J. for Sct. Res. Agri. Sc. 1999;**4**(2):19-25

[118] Ahmad F, Hafiz IA, Isi AA, Ahmed S, Khan M. Mango varietal susceptibility to malformation and its control. Asian Journal of Plant Sciences. 2002;**1**(2):158-159

[119] Iqbal Z, Dasti AA, Saleem A. Role of *Fusarium mangiferae* in causation of mango malformation disease. Journal of Research (Science). 2006;**17**(1):9-14

[120] Ploetz RC. Malformation: A unique and important disease of mango, *Mangifera indica* L. In: Summerell BA, Leslie JF, Backhouse D, Bryden WL, Editors, Fusarium: Paul E. Nelson Memorial Symposium. The American Phytopathological Society, St. Paul, MN; 2001. pp. 233-247

[121] Ploetz RC, Zheng QI, Vazguez A, Abdel Sattar MA. Current status and impact of mango malformation in Egypt. International Journal of Pest Management. 2002;**48**:279-285

[122] Noriega-canto DH, Telliz D, Mora-Aguilera G, Rodriguez-Alcazar J, Zavaleta-Mejfa E, Otero-Collinas G, et al. Epidemiology of mango malformation in Guerrero, Mexico, with traditional and integrated management. Plant Disease. 1999;**83**:223-228

[123] Britz H, Steenkamp ET, CoutinhoTA, Wingfield BD, Marasas WF, Wingfield MJ. Two new species of Fusarium section Liseola associated with mango malformation. Mycologia. 2002;**94**(4):722-730

[124] Lonsdale JH, Kotze JM. Chemical control of mango blossom disease and the effect on fruit set and yield. Plant Disease. 1993;**77**:558-562

[125] Kumar P, Chakrabarti DK, Dasgupta MK, Ghosh DC, Dasgupta D, Majumdar DK, et al. Mango malformation: Effect of mangiferin on morphology and parasitism in *Fusarium moniliforme*. Proceedings of National Symposium on Sustainable in Sub-Humid Zone. 1995;**3-5**:348-352

[126] Iqbal Z, Pervez MA, Ahmad S, Iftikhar Y, Yasin M, Nawaz A, et al. Determination of minimum inhibitory concentrations of fungicides against fungus Fusarium mangiferae. Pakistan Journal of Botany. 2010;**42**(5):3525-3532

[127] Freeman S, Shitenberg D, Maymon M. New insights into mango malformation disease epidemiology lead to a new integrated management strategy for sub-tropical environments. The American Phytopathological Society. Plant Disease. 2014;**98**:11

[128] Hafeez-ur-Rahman KMA, Khokhar KM, Shah AH, Msahib-ud-din K. Effect of benlate spray, biological antagonists and time of deblossoming on malformation of mango inflorescence. Pakistan Journal of Agricultural Research. 1989;**10**(1):60-65

[129] Rymbai H, Rajesh AM. Mango malformation: A review. Life sciences Leaflets. 2011;**22**:1079-1095

[130] Avneet K, Nirmaljit K. Mango malformation: A fungal disease, physiological disorder or malady of stress. Journal of Applied and Natural Science. 2018;**10**(1):403-409

[131] Ploetz RC. The major diseases of mango: Strategies and potential for sustainable management. Acta Horticulturae. 2004;(645):137-150

[132] Kumar P, Misra AK, Modi DR. Current status of mango malformation in India. Asian Journal of Plant Sciences. 2011;**10**(1):1-23

[133] Khaskheli MI, Pathan MA, Jiskani MM, Abro MA, Poussio GB, Khaskheli AJ. Effectiveness of different fungicides against predominant and virulent fungus *Fusarium nivale* the cause of mango malformation disease. Pakistan Journal of Phytopathology. 2017;**29**(01):137-143

[134] Khalid PA, Alam SS. Assessment keys for some important diseases of mango. Pakistan Journal of Biological Sciences. 2002;**5**:246-250

[135] Jiskani MM. Mango Diseases and their Management. Available from:

http://www.pakissan.com/english/advisory/mango.diseases.and.their.management.shtml [Accessed on: 2018-06-12]

[136] Rebolledo-Martínez A, del Angel-Pérez AL, Peralta-Antonio N, Díaz-Padilla G. Sooty mold control (*Capnodium mangiferae* Cooke and Brown) with biofungicides in leaves and fruits of mango "Manila" Tropical and Subtropical Agroecosystems. 2013;**16**:355-362

[137] Ahmed I, Mahmood A, Majeed M, Saleem A. Evaluation of various fungicides against die-back disease caused by *Diplodia natalensis* in mango. Pakistan Journal of Phytopathology. 1995;7:208-209

[138] Ploetz RC, Prakash O. Foliar, fl oral and soilborne diseases of mango. In: Litz RE, editor. The Mango: Botany, Production and Uses. Wallingford, UK: CAB International; 1997. pp. 281-325

[139] Shahbaz M, Khan SM, Iqbal Z, Rehman A, Muhammad F, Saleem A. Etiological studies to explore the causal agent of mango decline in the Punjab, Pakistan. Pak. Journal of Phytopathology. 2005;**17**:33-35

[140] Leghari TN. Epidemiology and yield losses by sudden death syndrome in mango orchards of Sindh and its possible control [M. Sc. Thesis]. Pakistan: Department of Plant Pathology, SAU Tandojam; 2005

[141] Hakro NH. Plant parasitic nematodes associated with mango orchards and their interaction with other soil borne pathogens [M.Sc. Thesis]. Pakistan: Department of Plant Pathology, SAU Tandojam; 2006

[142] Khanzada MA, Lodhi AM, Saleem S. Mango decline in Sindh: etiology and control. Souv. 5th meeting Intl. Symp. on Recent trends in plant disease management; 20-22 Dec., 2005. p. 78

[143] Saleem S, Khanzada MA, Lodhi AM. Mango Decline in Sindh. Abstracts, 9th Natl. Conf. of Pl. Scientist; 13-15 Feb. 2006. p. 28

[144] Asad M, Shafqat S, Da Silveira SF, Akem CN, Nazim H, Muhammad F. Quick decline of mango in Pakistan: Survey and pathogenicity of fungi isolated from mango tree and bark beetle. Pakistan Journal of Botany. 2011;**43**(3):1793-1798

[145] Das G, Zachariah AT. Dieback of mango, a new disease in India. Indian Journal of Botanical Science. 1945;**24**:101-108

[146] McSorley R, Campbell CW, Goldweber S. Observations of mango decline in South Florida. Proceedings of the Florida State Horticultural Society. 1981;**93**:132-133

[147] Narasimhudu Y, Reddy PSN. A note on gummosis of mango. Indian Phytopathology. 1992;**45**:261-262

[148] Sharma I. A note on population dynamics and etiology of die back of mango in Himachal Pradesh. The New Agriculturist. 1993;**2**(2):229-230

[149] Schaffer B. Mango Decline. In: Ploetz RC, Zentmyer GA, Nishijima W, Rohrbach HD, editors. Compendium of Tropical Fruit Diseases. St. Paul: APS Press; 1994. p. 43

[150] Simone GW. Disease Control in Mango (*Mangifera indica*). Plant Pathology Department Document PDMG-V3-22. Florida Cooperative Extension Service, Institute of Food and Agricultural Sciences, University of Florida; 1999

[151] Gonztilez E, Umana G, Arauz LF. Population fluctuation of

Botryodiplodia theobromae in mango. Agronomia Costarricense. 1999;**23**:21-29

[152] Al Adawi AO, Deadman ML, Al Rawahi AK, Khan AJ, Al Maqbali YM. *Diplodia theobromae* associated with sudden decline of mango in the Sultanate of Oman. Plant Pathology. 2003;**52**:419-419

[153] Ribeiro IJA. Seca de manguera. Agentes causais e studio da molesta. In: Anais do I Simposio Brasiliero Sobre a Cultura de Mangeura. Jacoticobal: Sociedad Brasileira de Fruticultura; 1980. pp. 123-130

[154] Malik MT, Dasti AA, Khan SM. Mango decline disorders prevailing in Pakistan. In: Malik AU, Pervez MA, Zaif K, editors. Proc International Conference on Mango and Date Palm, Culture and Export. June: 20-23; Faisalabad: University of Agriculture; 2005

[155] Saeed S, Masood A. Association of Bark beetle *Hypocrphalus mangiferae* Stebbing (Coleptera: Scolytidae) with pathogens *Ceratocystis fimbriata* and *Phompsis sp.,* in relation to Mango Sudden Death in Pakistan. International conference, 93rd ESA Annual Meeting, Milwaukee, Wisconsin August 3-8, 2008; USA

[156] Masood A, Saeed S, Erbilgin N, Kwon YJ. Role of stressed mango host conditions in attraction of and colonization by the mango bark beetle *Hypocryphalus mangiferae* Stebbing (Coleoptera: Curculionidae: Scolytinae) and in the symptom development of quick decline of mango trees in Pakistan. Entomological Research. 2010;**40**:316-327

[157] Galdino TVDS, Kumar S, Oliveira LSS, Alfenas AC, Neven LG, Al-Sadi AM, et al. Mapping global potential risk of Mango sudden decline disease caused by *Ceratocystis fimbriata*. PLoS One. 2016;**11**(7):e0159450. DOI: 10.1371/journal.pone.0159450

[158] Van Wyk M, Al Adawi AO, Khan IA, Deadman ML, Al Jahwari AA, Wingfield BD, et al. *Ceratocystis manginecans* sp. nov., causal agent of a destructive mango wilt disease in Oman and Pakistan. Fungal Diversity. 2007;**27**:213-230

[159] Khaskheli MI, Jiskani MM, Soomro MH, Talpur MA, Poussio GB. Prevalence of mango sudden decline/death syndrome (MSDS) on various varieties at the orchards of different age in the vicinity of Tando Qaiser, Hyderabad, Sindh. Pakistan Journal of Agriculture, Agricultural Engineering and Veterinary Sciences. 2011;**27**(2):160-167

[160] Sahi ST, Habib A, Ghazanfar MU, Badar A. *In-vitro* evaluation of different fungicides and plant extracts against *Botryodiplodia theobromae*, the causal agent of quick decline of mango. Pak. Journal of Phytopathology. 2012;**24**(2):137-142

[161] Sutton BC. The Coelomycetes. Kew, Surrey, England: Commonw. Mycolog.l Inst; 1980

[162] Khanzada MA, Lodhi AM, Saleem S. Pathogenicity of *Lasiodiplodia theobromae* and *Fusarium solani* on mango. Pakistan Journal of Botany. 2004;**36**(1):181-189

[163] Talukdar MJ. Plant disease in Bangladesh. Bangladesh Journal of Agricultural Research. 1974;**1**:61-83

[164] Singh S, Pandey PC, Misra BM. Twig blight and pod blackening of cacao (*Theobroma cacao*). Indian Forester. 1977;**103**:483-485

[165] Ilag LL, Marfil VE. Diplodia pod rot of mungbean. Philippine Agriculturist. 1977;**61**:486-491

[166] Domsch KH, Gams W, Anderson TH. Compendium of Soil Fungi. NY: Academic Press; 1980

[167] Ahmed S, Iqbal SH, Khalid AN. Fungi of Pakistan. Lahore, Pakistan: Sultan Ahmed Mycolog. Soc. Pakistan, Dept. Bot., Univ. Punjab; 1997

[168] Rehman A u, Umar U u D, Syed Atif HN, Munaza RL, Sajid AK, Muhammad TM, et al. Emerging resistance against different fungicides in *Lasiodiplodia theobromae* as the cause of mango dieback in Pakistan. Archives of Biological Sciences (Belgrade). 2015;**67**(1):241-249. DOI: 10.2298/ABS140904030R

[169] Saleem S, Khanzada MA, Lodhi MA. Mango decline in Sindh. Abstracts, 9th Natl. Conf. of Pl. Scientists 13-15 Feb. 9NCPS-2-16; 2006. p. 28

[170] Naqvi SAH, Perveen R, Malik MT, Malik O, Umer UD, Wazeer MS, et al. Characterization of symptoms severity on various mango cultivars to quick decline of mango in district Multan. International Journal of Biosciences. 2014;**4**:157-163

[171] Khanzada MA, Lodhi AM, Rajput AQ, Syed RN, Shahzad S. Response of different mango cultivars to mango decline pathogen, *Lasiodiplodia theobromae*. Pat. International Journal of Biology and Biotechnology. 2015;**12**:643-647

[172] Naqvi SAH, Perveen R. Mango quick decline manifestation on various cultivars at plants of particular age in the vicinity of district Multan. Pak. Journal of Phytopathology. 2015;**27**:31-39

[173] Kazmi M, Fateh F, Majeed K, Kashkhely AM, Hussain I, Ahmad I, et al. Incidence and etiology of mango sudden death phenomenon in Pakistan. Pak. Journal of Phytopathology. 2005;**17**:154-158

[174] Paolinelli-Alfonso M, Villalobos-Escobedo JM, Rolshausen P, Herrera-Estrella A, Galindo-Sánchez C, López-Hernández JF, et al. Global transcriptional analysis suggests Lasiodiplodia theobromae pathogenicity factors involved in modulation of grapevine defensive response. BMC Genomics. 2016;**17**:615

[175] Esam Eldin S, Arjun S, Zarqa AA, Al Shurafa KA, Al Naqbi TS, Iratni R, et al. Detection and management of mango dieback disease in the United Arab Emirates. International Journal of Molecular Sciences. 2017;**18**(10):2086. DOI: 10.3390/ijms18102086

[176] Malik MT, Ammar M, Ranan M, Rehman A, Bally ISE. Chemical and cultural management of die back disease of mango in Pakistan. Acta Horticulturae. 2016;(1111):363-368. DOI: 10.17660/ActaHortic.2016.1111.52

[177] Al-Jabri MK, Al-Shaili M, Al-Hashmi M, Nasehi A, Al-Mahmooli IH, Al-Sadi AM. Characterization and evaluation of fungicide resistance among *Lasiodiplodia theobromae* isolates associated with mango dieback in Oman. Journal of Plant Pathology. 2017;**99**(3):753-759

Chapter 9

Use of Parasitoids as a Biocontrol Agent in the Neotropical Region: Challenges and Potential

Yelitza C. Colmenarez, Natália Corniani, Simone Mundstock Jahnke, Marcus Vinicius Sampaio and Carlos Vásquez

Abstract

The agricultural production in the Neotropical region is highly affected by the attack of pests and diseases. Due to the overuse of pesticides, sustainable methods of control are in demand, such as biological control. Integrated Pest Management (IPM) considered the use of Biological control as a method to suppress the population of pests in several field agricultural systems and in protected crops systems. Biological control is generally appreciated today as an important component of IPM, and the demand for it is likely to spread as the IPM programs develop worldwide. The tropics present an important region for the application of biological control. The Neotropical region is characterized by its rich biodiversity, resulting in a wide range of natural enemies of pests represented by parasitoids, predators, and pathogens. Parasitoids are the natural enemies most used around the world for biological control. In this chapter, we present biological control programs using parasitoids established in the Neotropical region to control key pests of economic importance. Agricultural practices that maintain and enhance the action of natural enemies in crops will be reviewed, as the challenges and potential for the establishment of Biological Control programs using parasitoids in the Neotropical region.

Keywords: integrated pest management, sustainable agriculture, biological control, parasitoids, agricultural pests

1. Introduction

The agricultural production in the Neotropical region is highly affected by the attack of pests and diseases. Due to the overuse of pesticides, sustainable pest control methods are in demand, within the context of Integrated Pest Management (IPM) [1]. Biological control is a tool used in Integrated Pest Management (IPM) for several field agricultural systems and in protected crops systems [2]. Biological control, the use of living organisms as pest control agents, has enjoyed varying popularity in the past, but today is well established as an important component of IPM. Biological control is the most environmentally safe and economically profitable pest management method, when considering all the different factors together and its benefits to them.

The Neotropical region is characterized by its rich biodiversity, providing the opportunity to use a wide range of natural enemies of pests represented by parasitoids, predators, and pathogens [3, 4].

Dozens of species of insect predators and parasitoids are reared worldwide, and in some instances, these programs have been shown to be economically competitive with alternative methods of control. According to [5] in 2010, no less than 230 species of invertebrate natural enemies—originating from 10 taxonomic groups—were used in pest management worldwide. From this, within the arthropods, 52.2% were represented by parasitoids of the Hymenoptera group. Parasitoids can be used as biological control agents against insect pests in agro-ecosystems. Today, parasitoids are the most used natural enemies for classical biocontrol around the world, and many success cases have been reported in many countries of the Neotropical Region [2].

The term parasitoid defines the behavior of host use that exists only in insects (**Table 1**). Most of the parasitoids hosts are other insects and the parasitoids could be the same size as the host. Parasitoids can develop on or within their host, and parasitoids larvae kill their hosts to complete their life cycle from egg to adult and only need to feed on a single host to reach adulthood. The adult form has a free life. For killing their hosts, the parasitoids are the most effective natural enemies for pest biological control [2].

As an important aspect, the host imposes certain restrictions on the development of the parasitoids. In addition to this, the physiology and behavior of the host while it lives are in benefit of the parasitoid that develops, and when necessary, it can control them. As a result, the parasitoid has the opportunity to regulate host physiology [6]. Several parasitoids exhibit predatory adult behavior [7], but this does not alter the evolutionary interrelationships between the developing parasitoid and the host. Thus, the developmental duration of a parasitoid as carnivorous or saprophyte is continuous in some species, while *Trichogramma* rapidly kills the host and feeds on the preserved tissues [8]. Another interesting case is represented by the endoparasitoid from the Braconidae family, *Microplitis croceipes* (Cresson), which completes its development and emerges leaving the living host although reproductively dead [9].

There are different types of parasitoids:

Primary parasitoid: Species that develops on nonparasitized hosts.

Hyperparasitoid (or secondary parasitoid): Parasitoid that develops in another parasitoid (it is a parasitoid of a parasitoid). There may be several levels of hyperparasitism.

Endoparasitoid: Parasitoid that develops inside the body of the host. The endoparasitoid can be solitary (when a single larva completes its development in a given host) or gregarious (when several larvae develop to maturity in a single host).

Specialized in choice of host	**Tend to be smaller than the host**
Only the female searches for the host to lay eggs	Parasitoids develop on only one host individual during the immature stages
Eggs are usually laid in, on, or near the host	The immature stages remain on or in the host and almost always kill the host

Table 1.
Characteristics of parasitoids.

Ectoparasitoid: Species that develops outside the body of the host (they feed by inserting the buccal parts through the integument of the host). Like the endoparasitoids, there are solitary and gregarious endoparasitoids.

Multiple parasitism: Situation in which more than one species of parasitoids occur within or on a single host. In many cases, only one individual survives, others succumb. In rare cases, as species of *Trichoghramma* (parasitoids of lepidoptera eggs), more than one species can complete its development in the egg.

Superparasitim: In this case, several individuals of a species of parasitoids can develop in a host. When superparasitism occurs with solitary endoparasites, mutual destruction of the physiological suppression of larvae or surplus eggs may result in survival of a dominant individual. In some cases, however, the host dies prematurely, before the surplus is eliminated, and all die.

Adelphoparasitism: Also called autoparasitism, case in which a parasitoid species is a parasite of itself. For example, in *Coccophagus scutellaris* (Dalman), the male is obligatory parasitoid of the female.

Kleptoparasitism: In this case, a parasitoid preferentially attacks hosts that are already parasitized by other species. The kleptoparasitoid is not a hyperparasitoid. A kleptoparasitoid depends on another parasitoid to increase its reproductive success. It can act by opportunities, that is, using the oviposition holes or search trails made by another parasitoid species to lead it to the host.

Heteromes: The male and female are parasitoids of different hosts.

Polyembryony: The adult places a single egg per host, which later divides into many cells, each one developing independently. Several embryos are formed from a parasitized egg. It is common in Encyrtidae and Braconidae. The parasitic nematode *Ageniaspis citricola* of the citrus miner, *Phyllocnistis citrella*, produces 2–10 individuals per parasitized egg.

There are parasitoids of eggs, larvae (or nymphs), pupae, and adults.

Different types of biological control can be found: natural, conservation, inoculative (=classical), and augmentative biological control.

a. Natural biological control is the reduction of pest organisms by their natural enemies. It takes place in all of the world's ecosystems without any human intervention and, in economic terms, is one of the biggest contributions of biological control to agriculture and sustainable ecosystems [10].

b. Conservation biological control consists of human actions that protect and stimulate the performance of naturally occurring natural enemies [11].

c. Inoculative biological control, the natural enemies are collected in an area where they are present (usually the area of origin of the pest) and then released in new areas where the pest was accidentally introduced. The aim is that the offspring of the released natural enemies increases the populations in significant way in order to cause the suppression of pest populations during many subsequent years. This type of biological control has been used most frequently against introduced pests, which are presumed to have arrived in a new area without their natural enemies.

d. Classical biological control, it was the first type of biological control practiced widely [12]. This type of biological control implies the introduction of a natural enemy, which is from an exotic origin to control a pest, which usually is exotic. There are very successful cases as a result of the establishment of a classical biological control Program worldwide.

e. In augmentative biological control, natural enemies are mass reared in biofactories for release in large numbers in order to obtain an immediate control of pests. In some areas of agriculture, such as fruit orchards, maize, cotton, sugarcane, soybean, vineyards, and greenhouses, it has been considered to be an environmentally and economically sound successful alternative to chemical pest control [13].

In some countries, natural, conservation, and inoculative biological control are generally implemented using public funding, whereas augmentative biological control is often a commercial activity due to the need mass production and large scale regular releases of natural enemies.

Inoculative biological control is predicted to be applied on 350 million hectares worldwide [14], and over the last 120 years, 165 pest species have been brought under long-term control [15]. According to [15], 170 species of invertebrate natural enemies are mass produced and sold for release in augmentative biological control of approximately 100 pest species on about 0.4% of land under cultivation.

In this chapter, we present biological control programs established in the Neotropical region to control pests of economic importance such as *Diatraea saccharalis* (Lepidoptera: Crambidae), *Diaphorina citri* (Hemiptera: Psyllidae), *Phyllocnistis citrella* (Lepidoptera: Gracillariidae), *Maconellicoccus hirsutus* (Hemiptera: Pseudococcidae), *Paracoccus marginatus* (Hemiptera: Pseudococcidae), *Aleurocanthus woglumi* Ashby (Homoptera: Aleyrodidae), and *Aulacaspis yasumatsui* (Hemiptera: Diaspididae). Agricultural practices that maintain and enhance the action of natural enemies in crops will be reviewed, as the challenges and potential for the establishment of Biological Control programs using parasitoids in the Neotropical region.

2. Examples of successful biocontrol programs with parasitoids in the Neotropical Region

2.1 *Diatraea saccharalis* (Lepidoptera: Crambidae)

The sugarcane borer, *Diatraea saccharalis* (Fabricius, 1794) (Lepidoptera: Crambidae), is considered the most important sugarcane pest in the western hemisphere [16]. It is an insect pest of great economic importance in the sugarcane crop, due to the serious damages caused by its attack, which contributes to significant reductions in productivity and industrial use [17].

In Brazil, a very large program has been operating for 40 years to control the sugar cane borer with the larval parasitoid *Cotesia flavipes* Cameron (Hymenoptera: Braconidae), a braconid originally from Japan [18]. This is the most efficient biological program in Brazil, and it is among the best in the world. *C. flavipes* is released using inundative application in more than 3.3 million ha each year [19] (**Figure 1**). Another natural enemy used for this pest is the egg parasitoid *Trichogramma galloi*. In 2010, *T. galloi* was also used on 500,000 ha of sugarcane to control the eggs of the sugar cane borer [20].

In Colombia, sugar cane is the second most valuable crop. More than half of the surface area of the Cauca River Valley is planted to sugarcane, and the sugar cane borer has long been the principal pest causing high annual losses [21].

Efforts to improve biological control of *Diatraea* in Colombia began in the early 1970s with releases of *Trichogramma* spp. parasitoids (Hymenoptera:

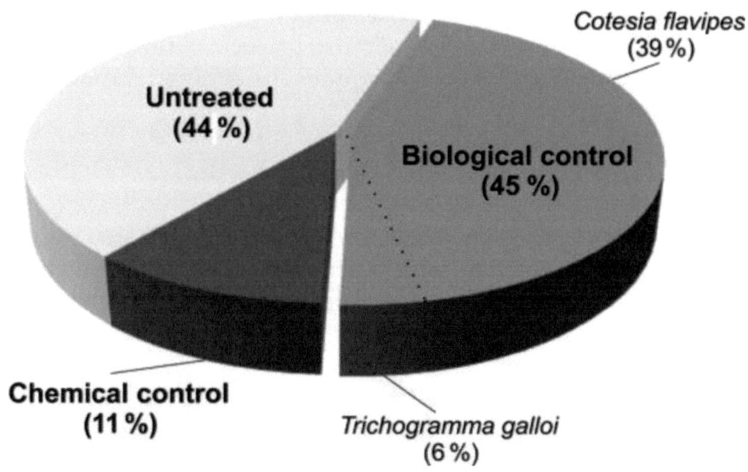

Figure 1.
Percentage of the sugarcane area in Brazil treated with releases of natural enemies [19].

Trichogrammatidae), followed by *Cotesia flavipes* that proved unsuited to conditions in the Cauca River Valley [22, 23]. Despite the introduction of several different geographic strains of *Cotesia flavipes* and repeated mass releases, few parasitoid cocoons were found in surveys, and it is not considered to be permanently established in the region [21].

Later, it was reported that only *Trichogramma exiguum* Pinto & Platner was recovered from eggs of three primary *Diatraea* species and the augmented species, *Trichogramma pretiosum* Riley, was not [23]. The development of integrated pest management (IPM) programs that incorporated the economic impact of the pests, their population dynamics, improved sampling procedures, and alternative methods of control such as use of the native egg parasitoids *T. exiguum* contributed to improved management of stem borers and reductions in their economic impact in Colombia [24].

2.2 *Maconellicoccus hirsutus* (Hemiptera: Pseudococcidae)

The hibiscus mealybug (HMB), *Maconellicoccus hirsutus* Green (Hemiptera: Pseudococcidae), was first detected in the Caribbean on the island of Grenada in 1993 [25], after which it rapidly spread through countries in the Caribbean, becoming one of the most important pest species [26].

Following the limited success of physical and chemical measures to control the pest populations, regional biological control programs for *M. hirsutus* were initiated in 1996. These have been joint efforts involving national programs with assistance from regional and international organizations like FAO, CARDI, CABI Bioscience, MCA, the USDA Animal and Plant Health Inspection Service (APHIS), and INRA [27].

Two parasitoids have been introduced and released against HMB, and these are *Anagyrus kamali* Moursi imported from China and Hawaii and *Gyranusoidea indica* Shafee, Alam, and Agarwal (Hymenoptera: Encyrtidae) imported from Egypt [28]. A third parasitoid, *A. dactylopii*, was also considered but was not introduced [26]. On some Caribbean islands, inundative releases of *Cryptolaemus montrouzieri* Mulsant were employed to provide supplemental control of heavy mealybug populations until parasitoids could be established [28].

The biological control program against this pest has been a tremendous success and may open the doors for further cooperative projects in that domain, and some of the experiences gained may be directly applied to finding solutions to problems caused by future invasive pests.

2.3 *Aleurocanthus woglumi* Ashby (Homoptera: Aleyrodidae)

The citrus blackfly (CBF) *Aleurocanthus woglumi* Ashby (Homoptera: Aleyrodidae) is assumed to be a native of Southern Asia [2]. The first report of *A. woglumi* in the Western Hemisphere was made by Ashby in Jamaica in 1915 [29]. From here, the citrus blackfly invaded other countries and now is widely distributed in the Western Hemisphere.

Damage caused by the CBF results from feeding, particularly on new growth of the host plants and from excretion of honeydew on which sooty-mold fungus develops. Badly infested leaves wither and drop-off as a result of injury caused by extraction of the cell sap. Such defoliation weakens the tree, interferes with its normal development and fruiting, and makes it unsightly [30].

CBF has several natural enemies. According to the many authors' reports, the most effective agents for controlling CBF are the parasitic wasps, *Encarsia perplexa* Huang & Polaszek (Hymenoptera: Aphelinidae) Polaszek (= *E. opulenta* Silvestri, Misidentified), and *Amitus hesperidum* Silvestri (Hymenoptera: Platygastridae) [31]. In Central America and Caribbean islands, the biocontrol of CBF is an important success case. In all the infested countries, the introduction of these biological control agents resulted in rapid and effective control of the CBF [32, 33].

In the 1990s, there was a resurgence of CBF as a serious problem in several countries in the Caribbean including Dominica, Guyana, French Guyana, St Kitts and Nevis, and Trinidad and Tobago [34]. In response to the threat posed by the pest, a classical biological control program was set up [26, 35] and the introduction of *A. hesperidum* in combination with *E. perplexa* appeared to be the best strategy for management of the pest [34].

2.4 *Diaphorina citri* (Hemiptera: Lividae)

Diaphorina citri (Kuwayama) is a global pest of citrus that transmits the bacteria *Candidatus liberibacter* spp. associated with the disease Huanglongbing (HLB) or greening. HLB is widespread in almost all citrus-producing regions except the Mediterranean [36]. Infection of groves results in complete loss of productive capacity within 4 years, and young citrus trees never produce [37].

According to [38] cited by [36], natural enemies play an essential role in regulating the population of *D. citri* in the field, and the elimination of these control agents by intensive use of insecticides can increase the spread of the disease.

Two parasitoid species associated with the citrus psyllid are *Tamarixia radiata* (Hymenoptera: Eulophidae), an idiobiont ectoparasitoid, and *Diaphorencyrtus aligarhensis* (Hymenoptera: Encyrtidae), an endoparasitoid [36]. *T. radiata* is believed to be more efficient than *D. aligarhensis* in controlling *D. citri* [36]. The wasp has provided excellent control of the psyllid on Reunion Island and good results on the islands of Guadeloupe and Puerto Rico [39, 40], Mexico [41], and Brazil [42], although its performance in Florida has been mediocre [38].

T. radiata has been reported in Brazil, Puerto Rico, Venezuela, Argentina, Colombia, Cuba, Mexico, and in the state of Texas, USA, without the need for a previous introduction [36].

2.5 *Paracoccus marginatus* (Hemiptera: Pseudococcidae)

Paracoccus marginatus Williams and Granara de Willink (Hemiptera: Pseudococcidae), popularly known as papaya mealybug (PMB), is a small hemipteran found to attack several genera of host plants and causes damage to many economically important ornamental and crop plants [43]. The PMB is native to Mexico and/or Central and North America [44, 45] and was first described in 1992 [46]. Since its first description, PMB has spread to several Caribbean islands and Central and South America countries [44, 45, 47].

PMB infestations are typically observed as clusters of cotton-like masses on the above-ground portion of host plants and cause damages to various parts including the leaves, stems, flowers, and fruits. The insect sucks the sap by inserting its stylets into the epidermis of the leaf, fruit, and stem. While feeding, it injects a toxic substance into the leaves resulting in curling, crinkling, rosetting, twisting, and general leaf distortion. Heavy infestations are capable of rendering fruit inedible due to the buildup of thick white wax [48, 49].

Mealybugs are difficult to control because they live in protected areas such as cracks, crevices and under the bark of their host plants. Besides that, most of the stages are covered with thick waxy that difficult its control with conventional insecticides [48, 49].

Biological control was identified as a key component in the PMB integrated management [50]. In addition to predators, several parasitoids may attack PMB. A total of 22 natural enemies either occurring naturally or introduced were reported on PMB in different countries, and *Acerophagus papayae* Noyes and Schauff (Encyritidea) is considered as one of the most efficient parasitoids for the suppression of PMB in its native range [51] (**Figure 2**).

2.6 *Phyllocnistis citrella* (Lepidoptera: Gracillariidae)

The citrus leafminer (CLM), *Phyllocnistis citrella* Stainton (Lepidoptera: Gracillariidae), is a pest native of Eastern and Southern Asia that since 1993 invaded all citrus-growing regions in America and the Mediterranean basin. The female moth lays its eggs on developing leaves, and the larvae form serpentine mines on the upper and lower surfaces of the leaves, sometimes even on the fruit. The CLM attacks and causes problems mainly in young trees, nurseries, and

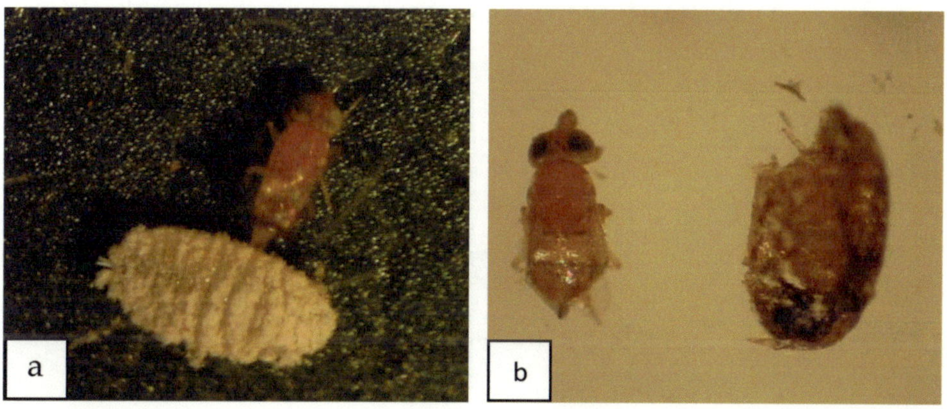

Figure 2.
a) Acerophagus papayae during parasitism process and b) Acerophagus papayae with its empty mummy with exit hole (Photos: Yelitza Colmenarez).

overgraftings [52]. Damages caused by the CLM include loss of photosynthetic capacity and stunting and malformation of leaves. In addition to the direct damage, the larval mining may facilitate the incidence of the citrus canker disease caused by *Xanthomonas axonopodis* pv. citri. Loss of access to international markets due to phytosanitary controls is a major economic impact related to leafminer damage [49].

The first attempt to control CLM in the tropics was using broad spectrum insecticides, but chemical control appeared to be a costly and short-term solution ([53, 54] in [52]).

According to [52], native parasitoids, in some environments, have been able to control CLM population, that is, *Galeopsomyia fausta* Lasalle, in Mexico, Central, and South America. In USA and Brazil, for example, the effort was also made in Classical Biological Control programs with the introduction of exotic parasitoids. In USA, an endoparasitic wasp of Asian origin, *Ageniaspis citricola* Logviniskaya (Hymenoptera: Eulophidae) was imported from Australia and released in Florida during 1994 and 1995 [55]. The population of *Ageniaspis citricola* quickly established and dispersed throughout the state, reaching parasitism levels near 100% in some areas [56].

In Barbados, *Ageniaspis citricola* was introduced in 2003–2004 from Florida and was released in the orchards around the island. Initial releases were conducted without taking into account the need of pruning to ensure new leaves and the presence of early larval stages (**Figure 3**). The biological control program was successful after the corrections of the initial problems. Currently, it is proving an excellent control of the pest. Another local parasitoid was found and identified as *Cirrospilus* sp. (Head of the Entomology Department Barbados Ministry of Agriculture Mr. Ian Gibbs).

Many native CLM parasitoids were identified in citrus groves in Brazil [57]. However, based on the parasitism potential of *Ageniaspis citricola* and the successful biological control of CLM achieved in other countries, it was decided to introduce this parasitoid in 1998 [58] (**Figure 4**). *Ageniaspis citricola* was soon established in several Brazilian states and it became the most common CLM parasitoid [59, 60].

2.7 *Aulacaspis yasumatsui* (Hemiptera: Diaspididae)

Cycads, commonly called "sago palms", are highly desired by landscapers and homeowners because they are long-lived, require low maintenance, and are resistant to most pests [61]. However, many cycad species are facing extinction

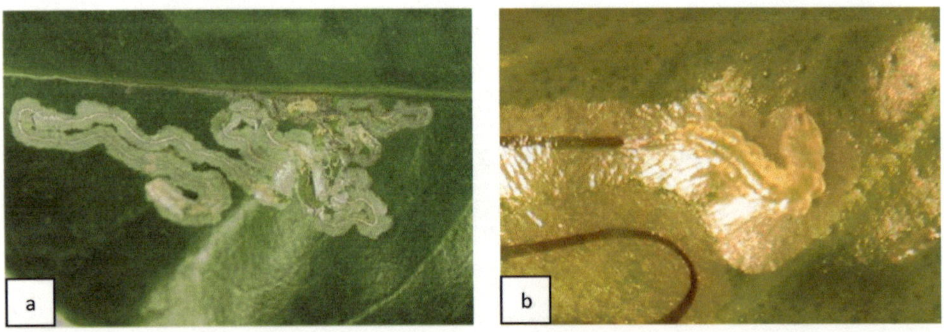

Figure 3.
a) Damage caused by the citrus leafminer. b) Early larval stage of citrus leafminer and its mine (photos: Yelitza Colmenarez).

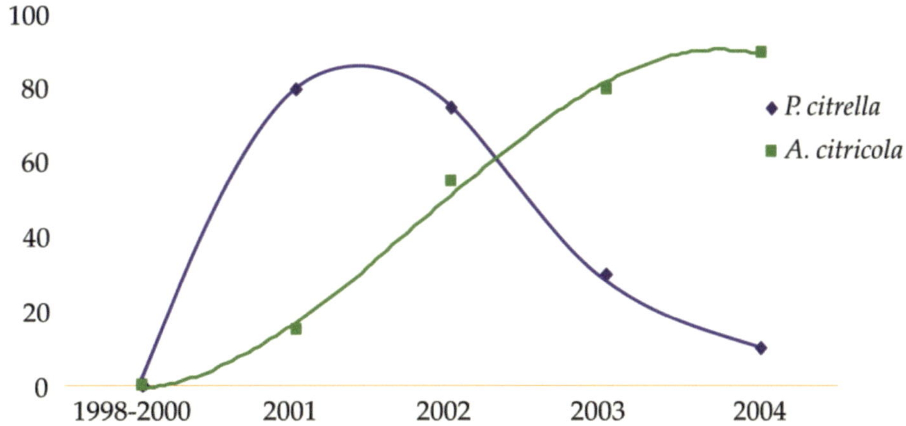

Figure 4.
Effect of Ageniaspis citricola on the Phyllocnistis citrella population [60].

in the wild due to insect pests [62]. The cycad aulacaspis scale (CAS), *Aulacaspis yasumatsui* Takagi (Hemiptera: Diaspididae), is one of the serious pest of cycads [63]. CAS was first described from species collected in Thailand [64]. The first report of CAS on cycads outside Thailand was in 1996 in Florida where it infested ornamental plants [65]; this pest was also reported in Mexico [66], Guam [67], the Cayman Islands, Puerto Rico, the Vieques Islands, and the Hawaiian Islands in the Americas [64].

This pest produces dense populations on the leaves, fruits, and trunk, resulting in premature death of leaves which can reduce plant longevity, as well as reducing its ornamental value. Many plants throughout Florida and the Caribbean have died as a result of this pest [61].

Chemical control can be expensive and provide inconsistent results [68]. Classical biological control of CAS began in 1998 when a parasitic wasp, *Coccobius fulvus* (Compere and Annecke), and a predatory beetle, *Cybocephalus nipponicus* Endrödy-Younga, were imported from Thailand and released in Florida [69]. In Barbados, both species were responsible for the suppression of the population of the pest in an effective way (Head of the Entomology Department Barbados Ministry of Agriculture Mr. Ian Gibbs). In addition, 16 species of predatory lady beetles (Coccinellidae) have been found on scale-infested plants in South Florida. However, nearly all these predatory natural enemies appear to be ineffective at providing satisfactory control. Therefore, research continues to examine natural enemies that may contribute to the overall biological control of the scale [68]. Besides, an alternative approach is to enforce strict quarantine measures in countries where CAS has not yet been introduced by prohibiting the importation of cycad plants from infested countries [64].

3. The increase of parasitoid action through conservative biological control

Conservative biological control is based on the preservation and/or modification of the environment through anthropic interventions to maintain and increase the survival of natural enemies in agroecosystems, in addition to improving their performance in natural pest control.

Interactions between species in an ecological system not only act on the species populations involved but also regulate the operation of complex networks.

These networks describe interactions between individuals or species and evaluate emerging system properties such as stability, resilience, or efficiency of energy transfer [70].

To reach success in conservative biological control, it is necessary to know the structure and function of each trophic level of a network in the environment, and so it can be managed in a way that the desirable species can be conserved in the system [71]. A basic point in this proposal is the diversification of vegetation in the cultivated area which favors natural enemies due to the availability and abundance of pollen and nectar that can meet their nutritional needs [72]. Offering refuge areas and appropriate microclimate, as well as alternative prey and host for development, are also important [73].

Habitat enhancement for natural enemies has been researched around the world through "habitat management" that has become a subdiscipline of pest management in many ecological studies centers [74].

About parasitoids, specifically, the studies have focused mainly on two axes: (1) The action of wild or native vegetation adjacent to the crop in the diversity (richness and abundance) of parasitoids and (2) and the increase of the parasitism with the diversification of the vegetation in polycultures or use of flower strip.

The first axis, the influence of wild vegetation on the diversity and abundance of parasitoids in cultivated areas, was evaluated in different regions and crops.

An example is the work of [75] who identified and compared diversity of parasitoid assemblages in an irrigated rice crop under organic management (OR) and in a nearby protected area (Wildlife Refuge Banhado dos Pachecos—BPWR) in the south of Brazil. Specimens were collected with Malaise and Moericke traps. As expected, the authors found a greater parasitoid diversity at the BPWR than at the OR. But the interesting thing is that the Platygastridae and Braconidae families, important natural enemies of agricultural pests, were the ones that had the highest number of morphospecies shared between the areas.

Thus, it is possible to infer that the legal reserve area of wild vegetation may be serving as a natural repository of parasitoid hymenoptera on organic crops if the distance between the areas is adequate.

To deepen the knowledge in this respect, another work was developed to evaluate the contribution of the presence of fragments of natural vegetation near rice-growing areas and the influence of different management of the crop on the abundance and diversity of families of hymenopteran parasitoids through distance gradients [76]. The work took place one rice crop with organic management (OM) and another one with conventional management (CM), in RS, Brazil, during two crop seasons. The parasitoids were collected with Malaise trap arranged at different distances in relation to the native vegetation surrounding the rice crop in both places. The most abundant families were Platygastridae, Mymaridae, Encyrtidae, Eulophidae, and Trichogrammatidae. Parasitoid average abundance was significantly higher on OM only in the second season. This may be due to the use of nonselective (neurotoxic) insecticides (neonicotinoids and pyrethroids) applied to the crop in 2014–2015 different from the first season analyzed, when growth regulators insecticides were used (buprofezin and benzylurea; farmer personal communication Mr. Denis).

There was a negative correlation between distance from native vegetation and parasitoid abundance in CM areas (**Figure 5**).

This result suggests the importance of this area in the presence of parasitoids in the crop. The role of these forest fragments in maintaining the richness and abundance of parasitoids has been described among others, by [77]. Interestingly, in the area of organic management, the distance gradient of the refuge did not significantly affect the abundance of parasitoids. The authors attribute this point to the

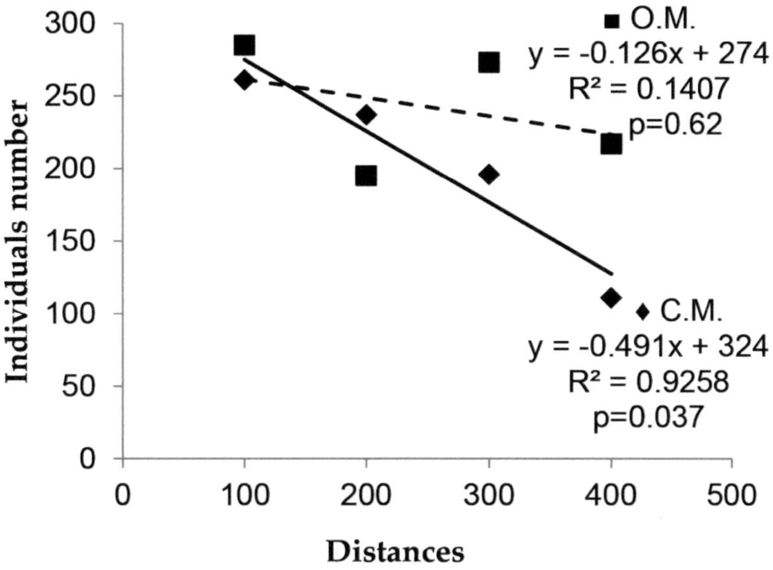

Figure 5.
Correlation between distances (Samples spots = 0, 100 m, 200 m, 300 m, and 400 m) from native vegetation (OM $p > 0.05$ and CM $p < 0.05$; Pearson Test) (first crop season) [76].

fact that in the OM area, there is a wild rice levee vegetation, including the presence of flowering species that increases the richness of parasitoids. The levees are created to produce a controlled flood environment in rice fields, and it has a variety of wild vegetation that growing in them. Therefore, the abundance may not be so dependent on the presence from preserved areas. Levees can serve as corridors extending the distances traveled by parasitoids [74].

The role of levee vegetation in the diversity of parasitoids was also tested in rice fields, comparing area [78]. One of them the wild vegetation from the levees was cut (C) monthly since the beginning of the planting period until the harvest, and in the other, the wild vegetation was not cut (NC). The average number of collected individuals by trap and sampling occasion was significantly larger in the NC area (42.5 ± 12.9) than in C area (23.9 ± 7.2; $H = 7.0687$; $p < 0.05$). The NC subarea has greater species richness than subarea C. This is demonstrated by the rarefaction curve which plotted the estimated number of morphospecies in relation to the number of individuals sampled (**Figure 6**). In the graph, the cutoff point (around 1200 individuals) shows that the richness curve observed in area C was below the 95% confidence limit of the curvae in the NC subarea, which had a larger parasitoid assemblage.

However, evaluating only egg parasitoids of plant hopper, [79] shows that these communities in rice fields are independent of the availability of noncrop habitats, providing additional nectar resources and retreat areas. The authors note that in contrast to temperate host parasitoid systems, rice hopper parasitoids seem to be very well adapted to the spatial and temporal heterogeneity because they have evolved in a rice monoculture system that offers sufficient resources that occur on traditional smallholder farms of Southeast Asia. Thus, other factors such as specific environmental and climatic conditions should be considered in the evaluation and implementation of conservative biological control.

Botanical species with high floral abundance can influence the attractiveness of floral visitors and natural enemies, since plants that offer more resources should be visited more frequently [80]. Therefore, the selection of plants with these characteristics is a relevant point.

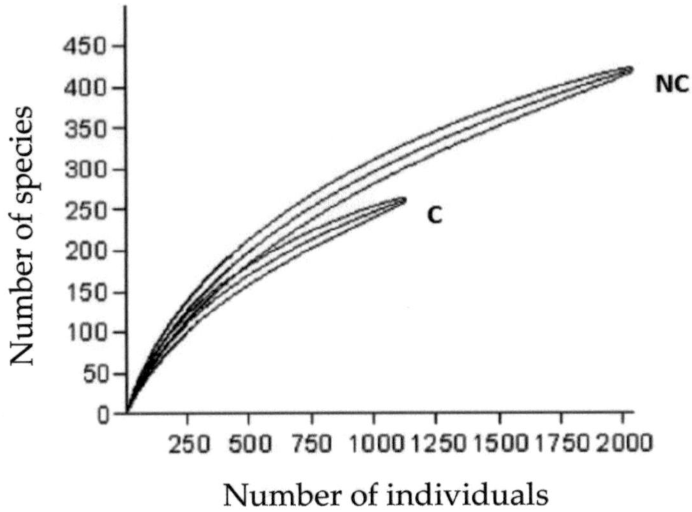

Figure 6.
Comparative richness rarefaction curves of hymenopteran parasitoids morphospecies collected in irrigated rice in organic system of production, in subareas not cut (NC) and cut (C) between October 2012 and March 2013 in Viamão, RS [78].

Nutritional quality of plants with flowers or extrafloral nectaries in the performance of the parasitoids has been studied in the laboratory and in the field, like the work of [81] who showed that the fertility and longevity of *Dolichogenidae tasmanica* (Cameron) (Hym., Braconidae) were increased in the treatments, in which there were flowers of *Lobularia maritima* (L.). [82], which also observed that adults of *Trichogramma carverae* Oatman & Pinto (Hym., Trichogrammatidae) had an increased survival rate with consequent increase in parasitism of *Epiphyas postvittana* (Walker) (Lep., Tortricidae) eggs, when confined with alyssum flowers.

In this way, we arrive at the second axis that focuses on the diversification of the vegetation in polyculture and the use of flower strips through habitat management. This kind of study can be exemplified by works such as [83], who concluded that tomato (*Solanum lycopersicum* L.) cultivated in polyculture with coriander (*Coriandrum sativum* L.) (Apiaceae), marigold (*Tagetes minuta* L.), and sorghum (*Sorghum bicolor* L.) presented lower losses due to pest attack. On the other hand, [84] stated that the wheat and pea consortium reduces the presence of pests not by increasing the number of natural enemies but by complicating the search of host plant for pests.

The distance between flowers is also an indirect factor that results in more attractiveness of beneficial insects; therefore, floral visitors should be more likely to visit flowers that are closer to each other, minimizing the energy spent in their activities [85]. In this sense, there are suggestions of flower arrangements between cultures such as the technical bulletin of Embrapa Agrobiologia [86] (**Figure 7**).

For example, the distance of the buckwheat flower strips grown between wheat (*Triticum aestivum* L. Poaceae) affected significantly the rate of the aphid *Metopolophium dirhodum* Walker (Hem., Aphididae) parasitism by the parasitoid *Aphidius rhopalosiphi* De Stefani-Peres (Hym., Aphidiidae), with exponential decline with increasing distance from flower strips [87].

Unfortunately, we do not have many examples of this kind of study in the American continent, but several countries in Europe have been studying the influence of flowers on survival, density, and pest control ability by parasitoids and predators. We can cite the work of [88], of the National Institute of Agricultural

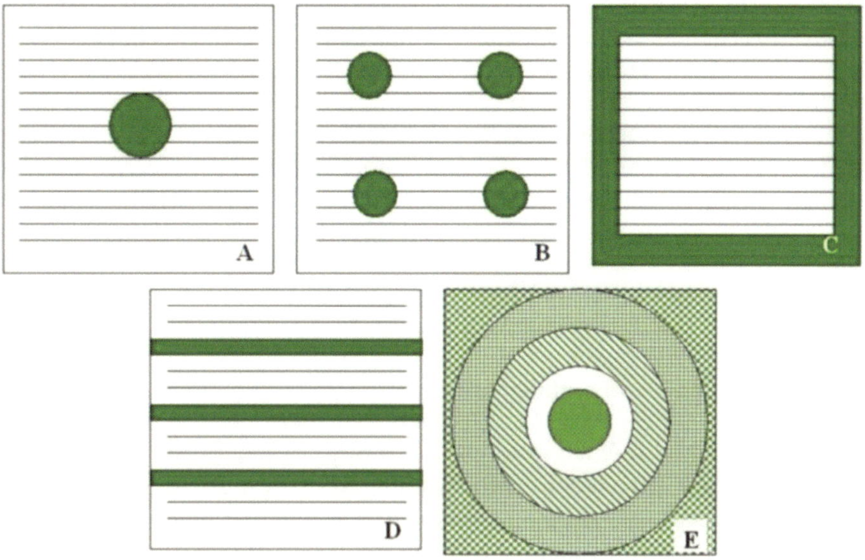

Figure 7.
Distribution arrangements of attractive plants (green) for natural enemies: central island (A); several islands (B); border (C); bands (C); mandala (D) [86].

Sciences (INCA) in Havana, Cuba who evaluated phytophagous insects and natural enemies in tomato-maize a polyculture and concluded that tomato-corn polyculture has a dissuasive influence on the development of pest populations in tomato crop besides enabling colonization by pest natural enemies.

In a work carried out at the Agricultural Research Company of Minas Gerais (EPAMIG), Brazil, the authors also evaluated the effect of the use of *Tagetes erecta* L. and *Calopogonium mucunoides* (Benth) on the occurrence of pest and natural enemies in crop of roses. The occurrence of *Praon volucre* (Hym., Braconidae) and *Pimpla croceiventris* (Cresson) (Hym., Ichneumonidae) was observed. The authors concluded that plant diversification contributed to a reduction in the occurrence of pests in this system [89].

Several interlayer designs of alders (*L. maritima*) were evaluated with organic lettuce for the control of aphids in ARS experimental fields in Salinas, California [90]. The authors also did studies with other plants, such as yellow mustard, and commented that these flowering plants attract syrphids and parasitoids that feed on pollen and flower nectar.

It is expected that this kind of study will increase in the Neotropical region in the next years with the increase in demand for alternative technologies of control and pest management.

Thus, we can reinforce the idea that multiple crop species grown in a single land increase biodiversity and encourage the presence of parasitoids. Plantation of multiple crops exploits different environmental niches, enhancing the total productivity per unit of land and expanding natural biological control.

4. Major challenges for the implementation of biological control programs in the Neotropical region

The list below highlights the major challenges that farmers, extension officers, and researchers have found when implementing biological control programs in Latin America and the Caribbean region:

a. Understanding and familiarization: At the farmer's level, it is important to ensure an efficient technology transfer method that facilitates the understanding of the technology and a high adoption at the field level. Experience working with farmers showed that once they understand and are familiar with the technology of application, way of action, and requirements of the bioproducts, they tend to use it much more. As an example of a good technology transfer methodology, the plant clinics established as part of the Plantwise program (www.plantwise.org) in the region can be mentioned. In this way, the extension officers/researchers provide information about sustainable methods of control, including the use of bioproducts, helping farmers to understand the way of action, and the best way/time for their application [1].

b. Integration of pests control methods: In order to change the current agricultural practices which depend heavily in the frequent application of pesticides, it is important to develop and transfer a package of sustainable production for the key crops. This will include the integration of methods of control, taking into account the crop phenology, time of the year when the pest attack most, and changes in climatic conditions per year that can favor the establishment and attack of new and current pests. In order to establish an effective Integrated Pest Management, it is important to highlight the need to study the selectivity of agrochemicals to natural enemies; this needs to be done case by case considering also local adaptations [1, 91].

c. Commercialization and availability of the bioproducts: Despite the high biodiversity the Neotropical region has and the potential for the use of Biological Control, the commercialization is a key factor to ensure the use of biological control agents at field level. Many potential/efficient biocontrol agents are not being commercialized as yet, among other reasons, in one hand due to problems when trying to mass produce them and in another hand due to their efficacy get compromised when integrating it with other methods of control. Therefore, more studies are needed to overcome these limitations, as well as the approximation and close work between scientific community and the companies that produce/commercialize the biocontrol products. Although some biological control agents are commercialized in countries in the region, it is very difficult for farmers to buy them in remote areas in some cases. Farmers in those locations can easily buy heavy toxic chemicals but their access to bioproducts seems to be very limited. Therefore, the distribution of these products in rural agricultural communities is also important to ensure their use [1, 91].

d. Government incentive: It is important that the government established strategies and incentives to ensure farmers, and in particular, small farmers located in very remote areas have the opportunity to get clear information about the bioproducts that are available in their region, understanding the way of actions, benefits, and minimal requirements to ensure their efficacy. In addition to this, clear information about the technology of application is crucial to ensure that the bioproducts show the level of efficacy expected. This can be achieved establishing a good and efficient national extension system [1].

5. Working group on parasitoids of the Neotropical region

In order to approximate scientists and share the information and results about the biological control programs carried out in the region, it is important to establish platforms/networks. The International Organization for Biological Control of Noxious Animals and Plants (IOBC—www.iobc-global.org) is represented in Latin America by the Neotropical Section (NTRS). The discussions about the importance of establishing the NTRS/IOBC started during the First Round Table for Biological Control in Santiago, Chile, in 1984. Following the propositions from this initial discussion, Jack Coulson (IOBC-Global President) and Jean-Paul Aeschlimann (IOBC-Global Secretary) contacted several biological control specialists in 1988 for the establishment of the Neotropical Regional Section (NTRS). In 1989, the NTRS was officially established during the 2nd Round Table for Biological Control held in Tucumán, Argentina. As part of the efforts of bringing scientist together, the working group on parasitoids of the Neotropical region was created in 2010. This working group is intended to improve the knowledge about parasitoids and their application in biological control in Latin America. The members of this working group share interests in all aspects of parasitoid lifestyle, such as taxonomy, diversity, ecology, behavior, and the utilization in biological control programs. This platform provides an opportunity for sharing information and joins efforts to promote parasitoids use as part of the biological control programs implemented in Latin America and the Caribbean [91].

6. Final considerations

The use of parasitoids in biological control programs has a big potential in the Neotropical region, which can be confirmed by the positive results obtained in the programs that are being implemented in the region. For example, the large areas under sugarcane cultivation in Brazil and to another extend in Colombia, using biological control of the sugarcane borer (*Diatraea saccharalis*) by the use of the larval parasitoid *Cotesia flavipes* and other species of parasitoids, make farmers and the agricultural community in general to perceive biological control as an efficient method that can be applied in large areas, increasing their interest in its use. In order to overcome the major challenges and constraints in parasitoids utilization as biological control agents in the region, it is necessary to grant farmers a better understanding of the benefits/requirements of these natural enemies. Helping farmers to become more familiar with the use of natural enemies, allowing an efficient integration of biological control with other control methods, could improve its application. It is worth mentioning that the commercialization and availability of the natural enemies are key factor to improve the adoption of biological control by farmers. Further studies are needed to develop and improve the mass production of some parasitoids species, which have been reported as efficient natural enemies of key pests.

Likewise, it is very important that the scientific community works together with extension officers to understand the major barriers experienced by farmers using biological control tactics, in order to present efficient, feasible, and cost effective solutions at field level. It is also important to reinforce the current platforms/networks where scientist, extension officers, students, industries, and practitioners could share results, experiences, and information to promote and incentive the use of biological control in a higher level in the region.

The implementation of biological control as part of an integrated pest management program could increase the sustainability of the agricultural production, reinforcing food security in the Neotropical region.

Author details

Yelitza C. Colmenarez[1*], Natália Corniani[1], Simone Mundstock Jahnke[2], Marcus Vinicius Sampaio[3] and Carlos Vásquez[4]

1 CABI South America, Botucatu, Brazil

2 Federal University of Rio Grande do Sul, Porto Alegre, Brazil

3 Federal University of Uberlândia, Uberlândia, Brazil

4 Technical University of Ambato, Ambato, Ecuador

*Address all correspondence to: y.colmenarez@cabi.org

References

[1] Colmenarez Y. In: Halfed-Vieira BA, Marinho-Prado JS, Nechet KL, Morandi MAB, Bettiol W, editors. Defensivos Agrícolas Naturais: Usos e Perspectivas. Brasília: Embrapa; 2016. pp. 802-843

[2] Sampaio MV, Bueno V, Silveira LCP, Auad AM. Biological control of insect pests in the tropics. In: Del Claro K, Oliveira PS, Rico-Gray V, editors. Tropical Biology and Conservation Management. Vol. 3. Oxford: UNESCO Publishing-Eolss; 2010. pp. 28-70

[3] Alves SB, Lopes RB, Pereira RM, Tamai MA. O controle microbiano na América Latina. In: Alves SB, Lopes RB, editors. Controle Microbiano de pragas na América Latina-Avanços e Desafios. Piracicaba: FEALQ; 2008. pp. 21-48

[4] Smith L, Bellotti AC. Successful biocontrol projects with emphasis on the Neotropics [Internet]; 1996. Available from: http://web.entomology.cornell.edu/shelton/cornell-biocontrol-conf/talks/bellotti.html [Accessed: May 17, 2018]

[5] van Lenteren JC. The state of commercial augmentative biological control: Plenty of natural enemies, but a frustrating lack of uptake. BioControl. 2012;**58**:1-20

[6] Vinson SB, Iwantsch GF. Host suitability for insect parasitoids. Annual Review of Entomology. 1980;**25**:397-419

[7] Chan MS, Godfray HCT. Host feeding strategies of parasitoids wasps. Evolutionary Ecology. 1993;**7**:593-604

[8] Vinson SB. Physiological interactions between egg parasitoids and their hosts. In: wajnberg E, Hassan SA, editors. Biological control with eggs parasitoids. Wallingford: Cab International; 1994. pp. 201-217

[9] Lewis WJ. Life history and anatomy of *Microplitis croceipes* (Hymenoptera: Braconidae) a parasite of *Heliothis* spp. (Lepidoptera: Noctuidae). Annual Entomology Society of America. 1970;**63**:67-70

[10] Waage JK, Greathead DJ. Biological control: Challenges and opportunities. Philosophical Transactions of the Royal Society of London B: Biological Sciences. 1988;**318**:111-112

[11] Gurr G, Wratten SD. Biological Control: Measures of Success. Vol. 429. Dordrecht, The Netherlands: Kluwer Academic Publishers; 2000

[12] DeBach P. Biological control of insect pests and weeds. New York: Reihold; 1964. 844 p

[13] van Lenteren JC, Bueno VHP. Augmentative biological control of arthropods in Latin America. BioControl. 2003;**48**:123-139

[14] Bale JS, van Lenteren JC, Bigler F. Biological control and sustainable food production. Philosophical Transactions of the Royal Society, B: Biological Sciences. 2008;**363**:761-776. DOI: 10.1098/rstb.2007.2182

[15] Cock JM, Sterck L, Rouzé P, Scornet D, Allen AE, Amoutzias G, et al. The *Ectocarpus* genome and the independent evolution of multicellularity in brown algae. Nature. 2010;**465**:617-621

[16] Capinera JL. Sugarcane borer, *Diatraea saccharalis* (Fabricius) (Insecta: Lepidopteera: Pyralidae). Gainesville: University of Florida; Entomology and Nematology Department; Florida Cooperative Extension Service; 2016

[17] Dinardo- Miranda LL, Pivetta JP, Fracasso JV. Influência da época de aplicação de nematicidas em soqueiras sobre as populações de nematóides e a produtividade da cana-de-açúcar. Bragantia. 2008;**64**:179-190

[18] Bolckmans K. Biocontrol: A quick global tour. In: Biocontrol files: Canada's bulletin on ecological pest management. Vol. 13. Canada: Agriculture and Agri-Food Canada; 2008. pp. 1-10

[19] Parra JRP. Biological control in Brazil: An overview. Scientia Agricola. 2014;**71**:345-355. DOI: 10.1590/0103-9016-2014-0167

[20] Parra JRP, Botelho PSM, Pinto AS. Biological control of pests as a key component for sustainable sugarcane production. In: Cortez LAB, editor. Sugarcane bioethanol: R&D for productivity and sustainability. São Paulo: Blücher; 2010. pp. 441-450

[21] Ocampo JAG. Memoria 1992-1993. Ministerio de Agricultura, Republic of Colombia, Santafé de Bogotá, Colombia; 1993

[22] Gómez LA, Lastra LA. Insects associated with sugarcane in Colombia. In: Cassalett C, Torres J, Isaacs C, editors. Sugarcane Cultivation in the Sugar-Producing Zone of Colombia. Cali, Colombia: Cenicaña; 1995. pp. 237-262

[23] Vargas G, Gómez LA, Michaud JP. Sugarcane stem borers of the Colombian Cauca River Valley: Current Pest Status, Biology, and Control. Florida Entomologist. 2015;**98**:728-735. DOI: 10.1653/024.098.0249

[24] Bustillo AE. Steps to reduce populations of Diatraea [Internet]; 2009. Available from: http://www.cenicana.org/publicaciones/carta_trimestral/ct2009/ct3y4_09/ct3y4_09_p10-15.php [Accessed: May 17, 2018]

[25] Persad C. Preliminary list of host plants of *Maconellicoccus hirsutus* Green: Hibiscus or pink mealybug in Grenada. Grenada: Caribbean Agricultural Research and Development Institute; 1995

[26] Kairo MTK, Pollard GV, Peterkin DD, Lopez VF. Biological control of the hibiscus mealybug, *Maconellicoccus hirsutus* Green (Hemiptera: Pseudococcidae) in the Caribbean. Integrated Pest Management Reviews. 2000;**5**:241-254

[27] Sagarra LA, Peterkin DD. Invasion of the Caribbean by the hibiscus mealybug, *Maconellicoccus hirsutus* Green [Homoptera: Pesudococcidae]. Phytoprotection. 1999;**80**:103-113

[28] Michaud JP. Three targets of classical biological control in the Caribbean: success, contribution, and failure. In: Van Driesche RG, editor. Proceedings of the First International Symposium on Biological Control of Arthropods. 2003. pp. 335-342

[29] Ashby SF. Notes on diseases of cultivated crops observed in 1913-1914. Bulletin of the Botanical Department, Jamaica. 1915;**2**:299-327

[30] Nguyen R, Hamon AB. Featured creatures [Internet]. 2016. Available from: http://entnemdept.ufl.edu/creatures/citrus/citrus_blackfly.htm [Accessed: May 14, 2018]

[31] Hart F, Henn W, Sontag H. Centro commercilae mondiale a New York city. In: Multy-storey buildings in Steel. London: Granada Publishin; 1978. pp. 162-164

[32] Cock MWJ. A review of biological control of pests in the commonwealth Caribbean and Bermuda up to 1982. In: Commonwealth Institute of Biological Control. Technical Communication. London: Commonwealth Agricultural Bureau; 1985 218 p

[33] Michaud JP. Biological control of Asian citrus psyllid, *Diaphorina citri* (Hemiptera: Psyllidae) in Florida: A preliminary report. Entomological News. 2002;**313**:216-222

[34] White GL, Kairo MTK, Lopez V. Classical biological control of the citrus blackfly *Aleurocanthus woglumi* by *Amitus hesperidum* in Trinidad. BioControl. 2005;**50**:751-759

[35] Persad R. Agricultural science for the Caribbean. Book 2. Cheltenham (UK): Nelson Thornes Ltd.; 2000

[36] Parra JRP, Alves GR, Diniz AJF, Vieira JM. *Tamarixia radiata* (Hymenoptera: Eulophidae) x *Diaphorina citri* (Hemiptera: Liviidae): Mass rearing and potential use of the parasitoid in Brazil. Journal of Integrated Pest Management. 2016;**5**:1-11

[37] Belasque Junior J, Yamamoto PT, Miranda MPD, Bassanezi RB, Ayres AJ, JM BOVÉ. Lessons from huanglongbing management in São Paulo State. Brazil. Journal of Plant Pathology. 2010;**92**:285-302

[38] Qureshi JA, Rogers ME, Hall DG, Stansly PA. Incidence of invasive *Diaphorina citri* (Hemiptera: Psyllidae) and its introduced parasitoid *Tamarixia radiata* (Hymenoptera: Eulophidae) in Florida citrus. Journal of Economic Entomology. 2009;**102**:247-256

[39] Etienne J, Quilici S, Marival D, Franck A. Biological control of *Diaphorina citri* (Hemiptera: Psyllidae) in Guadeloupe by imported *Tamarixia radiata* (Hymenoptera: Eulophidae). Fruits. 2001;**56**:307-315

[40] Pluke RWH, Qureshi JA, Stansly PA. Citrus flushing patterns, *Diaphorina citri* (Hemiptera: Psyllidae) populations and parasitism by *Tamarixia radiata* (Hymenoptera: Eulophidae) in Puerto Rico. Florida Entomologist. 2008;**91**:36-42

[41] Sanchez-Gonzalez JA, Sanchez-Borja M, Arredondo-Bernal HC. Cría masiva, liberación y evaluación en campo de *Tamarixia radiata* (Hymenoptera: Eulophidae). In: López-Arroyo JI, González-Lauck VW, editors. Memoria del Segundo Simposio Nacional sobre Investigación para el Manejo del Psílido Asiático de los Cítricos y el Huanglongbing en México. General Terán, Nuevo León, Mexico: INIFAP; 2011, 2011. pp. 339-344

[42] Gómez Torres ML. Estudos biológicos de *Tamarixia radiata* (Waterston, 1922) (Hymenoptera: Eulophidae) para o controle de *Diaphorina citri* Kuwayama, 1907 (Hemiptera: Psyllidae) [thesis]. Piracicaba: Escola Superior de Agricultura "Luiz de Queiroz", Universidade de São Paulo; 2009

[43] Mani M, Shivaraju C, Shylesha AN. *Paracoccus marginatus*, an invasive mealybug of papaya and its biological control: An Overview. Journal of Biological Control. 2012;**26**:201-216

[44] Miller DR, Williams DJ, Hamon AB. Notes on a new mealybug (Hemiptera:Coccoidea: Pseudococcidae) pest in Florida and the Caribbean: The papaya mealybug, *Paracoccus marginatus* Williams and Granada de Willink. Insecta Mundi. 1999;**13**:179-181

[45] Watson GW, Chandler LR. Identification of mealybugs important in the Caribbean Region. Commonwealth Science Council and CAB International, CARINET. 1999 40 pp

[46] Williams DJ, Granara De Willink MC. Mealybugs of Central and South America, CAB International, Wallingford, England; 1992. 635 p

[47] Kauffman WC, Meyerdirk DE, Warkentin R, Alvarez P, Abud A. Biological control of papaya mealybug in the Caribbean safeguarding the US Poster Presentation. International Organization for Biological Control, Bozeman, Montana, August 2-5, 2001

[48] Walker A, Hoy M. *Paracoccus marginatus* Williams and Granara de Willink Univ Florida Featured Creatures; 2018. Available from: http://entnemdept.ufl.edu/creatures/fruit/mealybugs/papaya_mealybug.htm [Accessed on: June 25, 2018]

[49] CAB International. Crop Protection Compendium. Wallingford, UK: CAB International; 2018

[50] Sankar C, Saravanan P, Kathiravan J, Marimuthu R, Prabhukumar S. Classical biological control of papaya mealybug, *Paracoccus marginatus* Williams and Granara de Willink (Hemiptera: Pseudococcidae) in tapioca by an encyrtide parasitoid, *Acerophagus papayae* in Perambalur District of Tamilnadu, India. Pest Management in Horticultural Ecosystems. 2012;**18**:213-216

[51] Amarasekare KG, Mannion CM, Epsky ND. Efficiency and establishment of the three introduced parasitoids of the mealybug, *Paracoccus marginatus* (Hemiptera:pseudococcidae). Biological Control. 2009;**51**:91-95

[52] Vercher R, Ferran MG, Comelles JC, Marzal C, Villalba M. Biological control of the citrus leafminer *Phyllocnistis citrella* (Lepidoptera: Gracillariidae) in Spain: Native parasitoids and establishment of *Citrostichus phyllocnistoides* (Hymenoptera: Eulophidae). Biological Control. 2004;**29**:215-226

[53] Knap JL, Albrigo LG, Browning HW, Bullock RC, Heppner JB, Hall DG, et al. Citrus leafminer, *Phyllocnistis citrella* Stainton: Current status in Florida. Florida, Gainesville: University of Florida. Florida Coop. Ext. Service. IFAS, Univ; 1995. p. 35

[54] Argov Y, Rössler Y. Introduction, release and recovery of several exotic natural enemies for biological control of the citrus leafminer, *Phyllocnistis citrella*, in Israel. Phytoparasitica. 1996;**24**:33-38

[55] Hoy MA, Nguyen R, Pomerinke M, Bullock R, Hall D, Knapp J, et al. Distribution of *A. citricola*–a parasitoid of the citrus leaf miner. Citrus Industry. 1997;**78**:51-52

[56] Walker A, Hoy M. A citrus leafminer parasitoid. University of Florida Featured Creatures. 2003. Available from: http://entnemdept.ufl.edu/creatures/beneficial/a_citricola.htm Accessed on: June 20, 2018

[57] Penteado-Dias AM, Gravena S, Paiva PEB, Pinto RA. Parasitóides de *Phyllocnistis citrella* (Stainton) (Lepidoptera: Gracillariidae: Phyllocnistinae) no Estado de São Paulo. Revista Laranja. 1997;**18**:79-84

[58] Paiva PEB, Yamamoto PT. Natural parasitism of the citrus leafminer (Lepidoptera:Gracillariidae) over eight years in seven citrus regions of São Paulo, Brazil. Florida Entomologist. 2015;**98**:660-664

[59] Sá LAN, Costa VA, Oliveira WP, Almeida GR. Parasitoids of *Phyllocnistis citrella* in Jaguariúna, State of São Paulo, Brazil, before and after the introduction of *Ageniaspis citricola*. Scientia Agricola. 2000;**57**:799-801

[60] Parra JRP. Sustainable pest control for Citrus production in Brazil. International Citrus Congress. 2016. Available from: http://www.icc2016.com/images/icc2016/downloads/

Jose_Parra_K3_20_09_2016.pdf Accessed: June 14, 2018

[61] Cave RD. Biological control agents of the cycad aulacaspis scale, *Aulacaspis yasumatsui*. Proceedings of the Florida State Horticultural Society. 2006;**119**:422-424

[62] Wu LW, Yen SH, Lees DC, Hsu YF. Elucidating genetic signatures of native and introduced populations of the Cycad Bule, Chilades pandava to Taiwan: A threat both to sago palm and to native cycas populations worldwide. Biological Invasions. 2010;**12**:2649-2669. DOI: 10.1007/s10530-009-9672-4

[63] Giorgi JA, Vandenberg NJ. Review of the lady beetle genus *Phaenochilus* Weise (Coleoptera: Coccinellidae: Chilocorini) with description a new species from Thailand that preys on cycad aulacaspis scale, *Aulacaspis yasum*atsui Takagi (Hemiptera: Sternorrhyncha: Diaspididae). Zootaxa. 2012;**3748**:239-255

[64] Wei J, Zhao Q, Zhao Q, Zhang H. Predicting the potential distributions of the invasive cycad scale *Aulacaspis yasumatsui* (Hemiptera: Diaspididae) under different climate change scenarios and the implications for management. Peer Journal. 2018;**6**:e4832. DOI: 10.7717/peerj.4832

[65] Howard FW, Hamon A, Mclaughlin M, Weissling T, Yang S. *Aulacaspis yasumatsui* (Hemiptera: Sternorrhyncha: Diaspididae), a scale pest of the cycads recently introduced to Florida. Florida Entomologist. 1999;**82**:14-27

[66] Normark BB, Normark RD, Vovides A, Lislie Solis-Montero L, Gonzalez-Gomez R, Pulido-Silva MT, et al. Cycad aulacaspis scale (*Aulacaspis yasumatsui* Takagi, 1977) in Mexico and Guatemala: A threat to native cycads. BioInvasions Records. 2017;**6**:187-193. DOI: 10.3391/bir.2017.6.3.02

[67] Terry I, Marler T. Paradise lost? Tipping the scales against Guam's Cycas micronesica. The Cycad Newsletter. 2005;**28**:21-23

[68] Cave R, Nguyen R, Manrique V, Avery PB. New research on two natural enemies of the cycad aulascaspis scale. The Cycad Newsletter. 2009;**32**:23

[69] Hodges G, Howard FW, Buss EA. Update on management methods for cycad aulacaspis scale. Division of Plant Industry: Florida Department of Agriculture and Consumer Services; 2003. Available from: http://www.doacs.state.fl.us/pi/enpp/ento/aulacaspis.html Accessed: June 22, 2018

[70] Pocok MJO, Evans DM, Fontaine C, Harvey M, Julliard R, McLaughlin O, et al. The visualisation of ecological networks, and their use as a tool for engagement, advocacy and management. Advances in Ecological Research. 2016. DOI: 10.1016/bs.aecr.2015.10.006

[71] Eilenberg J, Hajek A, Lomer C. Suggestions for unifying the terminology in biological control. Biological Control. 2001;**46**:387-400

[72] Roulston TH, Cane JH, Buchmann SL. What governs protein content of pollen: Pollinator preference, pollen pistil interactions, or phylogeny? Ecology. 2000;**70**:617-644

[73] Lavandero B, Wratten SD, Shishehbor P, Worner S. Enhancing the effectiveness of the parasitoid *Diadegma semiclausum* (Helen): movement after use of nectar in the field. Biological Control. 2005;**34**:152-158

[74] Gurr GM, Wratten SD, Landis DA, You M. Habitat management to

suppress pest populations: progress and prospects. Annual Review of Entomology. 2011;**62**:91-109

[75] Silva GS, Jahnke SM, Ferreira MLG. Hymenoptera parasitoids in protected area of Atlantic Forest biomes and organic rice field: Compared assemblages. Revista Colombiana de Entomologia. 2016;**42**:110-117

[76] Silva GS, Jahnke SM, Johnson N. Riparian forest fragments in rice fields under different management: differences on hymenopteran parasitoids diversity. Brazilian Journal of Biology. 2019

[77] Altieri M, Silva EN, Nicholls CI. O papel da biodiversidade no manejo de pragas. Holos; 2003. 226 p

[78] Simões-Pires PR, Jahnke SM, Readaelli LR. Influence of the vegetation management of the leeves in irrigated rice organic in diversity of Hymenoptera parasitoids. Brazilian Journal of Biology. 2016;**73**:774-781. DOI: 10.1590/1519-6984.06215

[79] Sann C, Theodorou P, Heong KL, Villareal S, Settele J, Vidal S, et al. Hopper parasitoids do not significantly benefit from non-crop habitats in rice production landscapes. Agriculture, Ecosystems and Environment. 2018;**254**: 224-232

[80] Carvalheiro LG, Biesmeijer JC, benadi G, Fründ J, Stang M, Bartomeus I, et al. The potential for indirect effects between co-flowering plants via shared pollinators depends on resource abundance, accessibility and relatedness. Ecology Letters. 2014;**17**:1389-1399

[81] Wratten SD, Lavandero BI, Tylianakis J, Vattala D, Çilgi T, Sedcole R. Effects of flowers on parasitoid longevity and fecundity. New Zealand Plant Protection. 2003;**56**:239-245

[82] Begum M, Gurr GM, Wratten SD, Hedberg P, Nicol HI. Using selective food plants to maximize biological control of vineyard pests. Journal of Applied Ecology. 2006;**43**:547-554

[83] Gomes FB, Fortunato LDJ, Pacheco ALV, Azevedo LH, Freitas N, Homma SK.Incidência de pragas e desempenho produtivo do tomateiro orgânico em monocultivo e policultivo. Horticultura Brasileira. 2012;**30**:756-761

[84] Lopes T, Bodson B, Francis F. Associations of wheat with pea can reduce aphid infestations. Neotropical Entomology. 2015;**44**:286-293

[85] Simba LD, Foord SH, Thébault E, van Veen FJF, Joseph GS, Seymour CL. Indirect interactions between crops and natural vegetation through flower visitors: The importance of temporal as well as spatial spillover. Agriculture, Ecosystems and Environment. 2018;**253**:148-156

[86] Aguiar-Menezes EL, Silva AC. Plantas atrativas para inimigos naturais e sua contribuição no controle biológico de pragas agrícolas. Brasília: Embrapa Agrobiologia; 2011 60 p

[87] Tylianakis JM, Klein AM, Tscharntke T. Spatiotemporal variation in the diversity of hymenoptera across a tropical habitat gradient. Ecology. 2004;**86**:3296-3302

[88] León A, Pino MDLA, González C, del Pozo E. Evaluación comparativa de densidades de fitófagos y enemigos naturales en policultivo tomate-maíz. Cultivos Tropicales. 2000;**21**:53-60

[89] Carvalho LM, Ladreira VA, Almeida EFA, Reis NA, Lessa MA. Efeito do uso de *Tagetes erecta* e *Calopogonium mucunoides* na ocorrência de pragas e inimigos naturais em cultivo de roseira. Resumos do VIII Congresso Brasileiro de Agroecologia Cadernos de Agroecologia,

Porto Alegre/RS. 2013;**8**. Available from: http://revistas.aba-agroecologia.org.br/index.php/cad/article/view/13807 Accessed: July 02, 2018

[90] Brennan EB. Agronomic aspects of strip intercropping lettuce with alyssum for biological control of aphids. Biological Control. 2013;**65**:302-311

[91] IOBC NTRS 2018. Available from: http://www.iobcntrs.org/GT.aspx [Accessed: June 02, 2018]

Chapter 10

Essential Oil Nanoformulations as a Novel Method for Insect Pest Control in Horticulture

Samar S. Ibrahim

Abstract

Eco-friendly biopesticides based on essential oils (EOs) appear to be a complementary or alternative method to chemically synthesized insecticides in integrated pest management programs. They have the advantage of reducing the adverse effects of chemical insecticides on human health and environment and at the same time increasing horticultural crop productivity and yield. Plant EOs exhibit toxic, repellent, and antifeedant effects on different insect species. However, the main problem in using plant EOs as biopesticides under field conditions is their chemical instability in the presence of air, light, moisture, and high temperatures which lead to the rapid evaporation and degradation of their active constituents. Incorporation of EOs into controlled-release nanoformulations may contribute to solve problems associated with their application; this kind of formulation is expected to be more effective than the bulk (free) substance.

Keywords: plant essential oils, biopesticides, nanoformulation, insect pest control, horticultural crops

1. Introduction

Horticultural crops are infested with numerous insect pests that cause tremendous economic losses including aphids (*Myzus persicae* and *Aphis gossypii*), beet armyworm (*Spodoptera exigua*), cabbage loopers (*Trichoplusia ni*), citrus mealybug (*Planococcus citri*), onion thrips (*Thrips tabaci*), and greenhouse whitefly (*Trialeurodes vaporariorum*) [1]. The use of synthetic chemical insecticides to control insect pests poses risks to human health and environment. For this reason, there is an urgent need to apply a range of modern strategies as alternatives for chemical pesticides in order to protect the environment from insecticidal pollution, decrease the regenerating resistance, and increase horticultural crop productivity.

Plant natural substances may provide potential alternatives to currently used insect-control agents because these materials constitute a rich source of bioactive chemicals. Plant-active substances may not only act as toxicants to insects but also as insect growth regulators, repellents, synergists, or phagodeterrents [2–4]. However, the major inconvenience of the use of essential oils is their chemical instability in the presence of air, light, moisture, and high temperatures that can lead to the rapid evaporation and degradation of some active components. Nanoformulations of the essential oils could solve these problems by protecting active components of essential oils from degradation and losses by evaporation, thereby enhancing their stability and maintaining the minimum effective dosage/application [5].

Nanoencapsulation or controlled delivery is a technique in which a membrane encloses small particles of active ingredient with the objective of offering protection to the core material from adverse environmental conditions, such as undesirable effects of light, moisture, and oxygen, and also avoiding drawbacks such as odor and volatility [6].

Nanoparticles (NPs) can be classified on the basis of the kind of material into metallic, semiconductor, and polymeric nanoparticles; the last ones are the most promising for essential oil nanoformulation. Furthermore, this kind of formulation is expected to be more effective than the bulk substances [7]. Controlled delivery technologies have emerged as an approach with promise not only to utilize resources in the maximum efficient way but also to reduce pollution. Moreover, if the resource is a natural or renewable polymer, then it will draw attention as a more new, more economical, and more eco-friendly source for use of humankind and a suitable approach for biological and integrated pest management (IPM) programs.

This chapter is describing the efficacy of essential oils as eco-friendly biopesticides and shedding light on the nanoformulations as biopesticides and their potential role in agriculture particularly in insect pest control. In order to reduce the negative impacts of chemical insecticides on environment and crop plants, and to protect crops from insect pests, nanoformulations are highly prospective to become an essential factor in integrated pest management programs.

2. Plant essential oils as biopesticides

Plants provide potential alternatives to currently used insect-control agents because they constitute a rich source of bioactive chemicals [8]; among these chemicals are plant essential oils. Essential oils, also known as aromatic oils, are volatile compounds produced naturally in plants for their own needs other than nutrition (i.e., protection or attraction) as secondary metabolites with distinctive odor [9, 10], most of them containing natural antioxidants and natural antimicrobial agents [11], and they are usually used in perfumery, in aromatherapy, in cosmetics, in incense, in medicine, in household cleaning products, and for flavoring food and drink [12].

Several essential oils have antiparasitic, bactericidal, fungicidal, virucidal, and insecticidal properties [2, 9]. Essential oils extracted from plants may act as toxicants, insect growth regulators [13], repellents and synergists [3, 14], and also as phagodeterrents [15]. Biopesticides based on essential oils (EOs) appear to be a complementary or alternative method in crop production and integrated pest management [16].

All chemicals produced by nature can be classified into two main groups; the first is the primary metabolites and constitutes the basic building blocks of living organisms such as proteins, carbohydrates, nucleic acids, and lipids. The second group is secondary metabolites that are simply classified into three main groups: terpenes (such as plant volatiles, cardiac glycosides, carotenoids, and sterols), phenolics (such as phenolic acids, coumarins, lignans, stilbenes, flavonoids, tannins, and lignin), and nitrogen-containing compounds (such as alkaloids and glucosinolates) [17]. Secondary metabolites play an important role in plant defense system against microorganisms and herbivorous insects [18].

According to [19] essential oils are lipophilic and thus can enter the insect and cause biochemical dysfunction and mortality. Several studies revealed that EOs from different plant families such as Asteraceae, Myrtaceae, Apiaceae, Lamiaceae, and Rutaceae are effective against different insect pests [19–22] (**Table 1**).

It has been shown that essential oils enter the insect body either by inhalation, ingestion, or skin absorption [23]. Essential oils may interfere with the biology, physiology, and nervous system of the insect [24, 25]. The biological and physiological effects of the botanical oils on insect pests may be attributed to their effect on the insect neuroendocrine system and juvenile hormone leading to hormone unbalance and insect malformation. A decrease in adult insect fecundity and egg fertility was observed [26] and may be caused by a decrease in periods and time of adults mating which leads to a reduction in ovulation [27]. Essential oils can also interfere with normal respiration of insects; it has been reported that essential oils can block the spiracles of insects leading to their suffocation [28–31].

Monoterpenes, the major components of plant essential oils, act as a neurotoxicant and act on acetylcholinesterase enzyme AChE, a key enzyme responsible for terminating the nerve impulse transmission through the synaptic pathway. Essential oil components also act on the octopaminergic system of insects, which is considered as a target for insect control. Insect paralysis and death

Plant family	Plant name	Insect pest	References
Tropaeolaceae	*Tropaeolum tuberosum*	*Premnotrypes vorax* (Coleoptera: Curculionidae)	[43]
Meliaceae Meliaceae Amaryllidaceae Anacardiaceae	*Melia azedarach* *Azadirachta indica* *Allium sativum* *Schinus molle*	*Phthorimaea operculella* (Lepidoptera: Gelechiidae)	[44]
Lamiaceae Rutaceae Tropaeolaceae Solanaceae Amaryllidaceae	*Minthostachys mollis* *Ruta graveolens* *Tropaeolum tuberosum* *Capsicum frutescens* *Allium cepa*	*Tecia solanivora* (Lepidoptera: Gelechiidae)	[45]
Poaceae	*Cymbopogon winterianus*	*Spodoptera frugiperda* (Lepidoptera: Noctuidae)	[46]
Myrtaceae Ericaceae	*Eucalyptus globulus* *Gaultheria procumbens*	*Agrotis ipsilon* (Lepidoptera: Noctuidae)	[47]
Amaryllidaceae	*Allium sativum*	*Tribolium castaneum* (Coleoptera: Tenebrionidae) *Sitophilus zeamais* (Coleoptera: Curculionidae)	[48]
Apiaceae Myristicaceae	*Athmanta haynaldii* *Myristica fragrans*	*Lymantria dispar* (Lepidoptera: Erebidae)	[49]
Poaceae	*Cymbopogon winterianus*	*Frankliniella schultzei* (Thysanoptera: Thripidae) *Myzus persicae* (Hemiptera: Aphididae)	[50]
Rutaceae	*Citrus sinensis*	*Ceratitis capitata* (Diptera: Tephritidae)	[51]
Myrtaceae Anacardiaceae	*Eucalyptus citriodora* *Schinus terebinthifolius*	*Bemisia tabaci* (Hemiptera: Aleyrodidae) *Trialeurodes ricini* (Hemiptera: Aleyrodidae)	[52]

Table 1.
Essential oils of some plant families with insecticidal efficacy against different insect pests.

predominantly occurred by disruption in octopamine, which is a neurotransmitter, neurohormone, and circulating neurohormone-neuromodulator, resulting in total breakdown of the nervous system [32–34].

The feeding efficiency is the ability of the insect species to use the food ingested to the best of their capabilities; antifeedant activity of the essential oils or extracts from different plants would be related to their effect on the chemoreceptors [35]. Plants containing alkaloids, steroids, flavonoids, terpenoids, and saponins possess high antifeedant activity against different insects; therefore these plants and different essential oils which exhibited antifeedant constituents could be developed into products suitable for using in integrated insect management programs [35]. The quality and quantity of food consumed may increasingly affect the growth, development, and reproduction of insects [36]; oil compounds can reduce ingestion or efficiency of conversion of assimilated materials and prevented nutrients from being available to own biomass [37, 38]. This positive effect of essential oils on insect-feeding efficiency may be attributed to the irreversible damage of some membranes related with the absorption in the gut; thus, large amount of energy is exerted by larvae to detoxify the essential oils [39].

Generally essential oils and their components have been considered safer than other plant-derived chemicals like rotenone and pyrethrum [40]. This could be attributed to existing detoxifying metabolism pathways and bio-rational mode of action of monoterpenoids as reported by [41]. It can be concluded that the essential oils act at multiple levels in the insects, so the possibility of generating resistance is a little probable [42] making them effective nontoxic agents in IPM programs.

3. Essential oil nanoformulations

Agricultural pests are usually controlled using chemical pesticides; 90% of applied pesticides are lost to the air and severely affecting the environment and increasing application costs to the farmer [53]. In addition, the use of pesticides increases pest resistance and reduces soil biodiversity [4, 53]. Recently, in many countries, integrated pest management systems which comprise both methods of traditional crop rotation and biological pest control are becoming favorable and preferred method to improve crop yields.

Biopesticides based on essential oils (EOs) appear to be a complementary or alternative method in IPM [54]. Essential oils showed toxic, repellent, and antifeedant effects on different insect species [2, 3, 55]. Despite these promising properties, problems related with the essential oils volatility, poor water solubility, and aptitude for oxidation have to be resolved before they can be used as an alternative pest control system [56].

Nanoemulsions and nanoencapsulation of the essential oils may solve these problems protecting them from degradation and losses by evaporation; this kind of formulation is expected to be more effective than the bulk (free) substance. On the other hand, it was found that pesticide nanoformulations showed less toxicity toward nontarget organisms compared with commercial formulations, and therefore a higher specificity was observed [57]; besides, that they reduce pesticide use and increase persistence of the active ingredient [58].

To achieve high stability and efficacy, essential oils are encapsulated in nanoemulsions, which are used as delivery systems and considered as a promising strategy to deliver essential oils in agriculture and particularly in insect pest management. Nanoemulsions are thermodynamically unstable systems that have small droplets (radius < 100 nm), which make them transparent or translucent [59], and could be used for both hydrophilic and hydrophobic pesticides; accordingly the use of toxic organic solvents can be eliminated. Nanoemulsion is usually produced either by high-energy emulsification or low-energy emulsification methods. In high-energy emulsification technique, mechanical devices such as high-pressure homogenizers, ultrasonic homogenizers, and microfluidizers are used to reduce droplet size by generating intense disruptive forces [60], while in low-energy emulsification technique, the physicochemical characteristics of surfactants and co-surfactants are involved [61]. It has been reported that using surfactant that blends in preparing nanoemulsions is usually more efficient than individual uses for various applications, using sufficient amounts of suitable surfactants and additionally protective colloids believed to make nanoemulsions more resistant to crystallization, agglomeration, and sedimentation [62].

Nanoencapsulation is a technique in which the active agent (solid, liquid, or gas) is surrounded or encapsulated by a thin layer or a membrane to protect the core active agent from severe and harmful environmental conditions such as light, moisture, and high temperature effects. The envelope or carrier could be natural polymers (polysaccharides, proteins), synthetic polymers (polyamides, melamine-formaldehyde, etc.), lipids, phospholipids, or inorganic materials (SiO_2) [63, 64].

Nanocarriers are structured in different designs with different materials; the main organic nanocarrier systems are polymeric and lipid-based particles. Polymeric nanoparticles are classified as nanocapsules (consist of polymeric wall and a core, which is commonly oily) and nanospheres (which are matrix systems) (**Figure 1**); these polymeric nanoparticles can be prepared using different techniques; one of them is known as nanoprecipitation or solvent displacement and based on an antisolvent procedure [65]. The other nanocarrier system is lipid nanoparticles which can be classified into liposomes, niosomes, solid lipid nanoparticles (SLNs), and nanostructured lipid carriers (NLC). To prepare these nanoparticles, several methods are used, for example, thin lipid film hydration method [66] and ethanol injection [67] to prepare liposomes, and solvent-in-water emulsion diffusion technique replacing liquid lipid of an o/w emulsion for a solid or a blend of solid lipids is used for the preparation of solid lipid nanoparticles [68].

A significant characteristic for nanoencapsulated system is controlled release, usually including an initial burst release, followed by a prolonged release [69]; obviously the main advantages of nanoencapsulation are its ability to reduce the amount of active ingredients needed, minimize evaporation, and control the release of active components. The use and application of nanoencapsulation in recent years

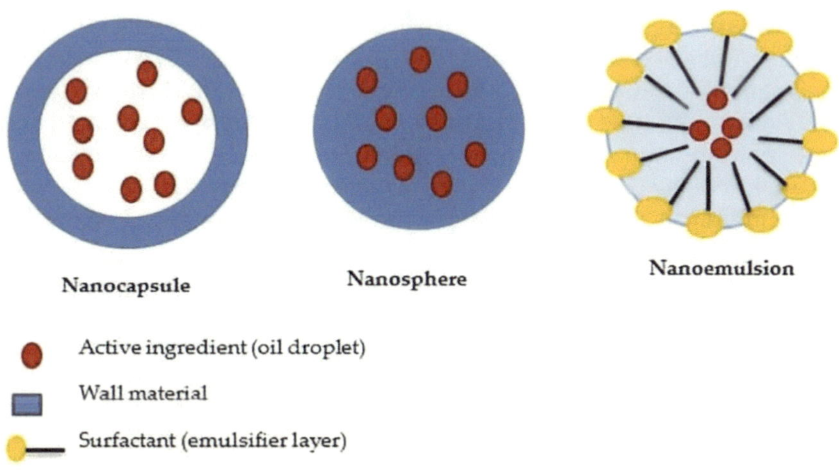

Figure 1.
Schematic representation of different nanoformulations.

have been increased. Manufactured nanoparticles exhibit a broad range of applications due to their unique properties compared with their bulk counterparts [70].

In prepared nanoencapsulated oil formulations, the encapsulation efficiency (% EE) can be expressed as a percentage of the total amount of oil found in the formulation at the end of the preparation procedure, or it is the ratio between the mass of entrapped essential oil and the total mass of the essential oil added, while the loading capacity (%LC) is the ratio between the mass of entrapped essential oil and the total mass of carrier (coating material) [71]. It was indicated by [72] that factors determining loading capacity and encapsulation efficiency are the solubility and miscibility of the active ingredient in the melted lipid phase, physiochemical structure of the lipid matrix, and the polymorphic state of the lipid material. A study by [71] mentioned that the potential of the particles depending on the nature of the coating nanoparticles also influenced the encapsulation efficiency and accordingly loading capacity. It has been reported that there are three variables that affect the encapsulation efficiency: stirring rate, oil loading, and the amount of cross-linking agent [73].

4. Essential oil nanoformulations and insect pest control

Nanoformulations exhibit unique properties compared with their bulk counterparts, including higher pest toxicity and less toxicity toward nontarget organisms. It was indicated that nanopermethrin has more larvicidal effect against *Culex quinquefasciatus* than the bulk form of permethrin [74]. Nanoformulation degrades rapidly with residual levels below the regulatory criteria in foodstuffs as concluded by [75] in their review dealing with applications of nanomaterials in agricultural production and crop protection.

Recent studies revealed the novel general and biological properties of the known materials, which they acquire when transformed into nanoparticles. They penetrate into the cells of the pest specially epithelial and endothelial cells by transcytosis. Moreover, they travel along the dendrites and axons, the blood, and lymphatic vessels provoking oxidative stress and other consequences [76].

In a previous study, geranium oil was incorporated into solid lipid nanoparticles using ultrasonic-solvent emulsification technique; the results indicated a production of high-quality solid lipid nanoparticle-loading geranium oil that was used as mosquito repellent [77]. While in other studies, polyethylene glycol (PEG) nanoparticles were used to incorporate garlic and geranium essential oils and were tested against *Tribolium castaneum* and *Rhyzopertha dominica* [16, 78], both essential oil-loaded nanoparticles produced a notable increase in the residual contact toxicity apparently due to the slow and persistence release of the active terpenes. In addition, the nanoformulation enhanced the essential oil contact toxicity and altered the nutritional physiology of both stored product pests. The essential oil citronella nanoemulsion prepared by high-pressure homogenization had resulted in a higher-release rate against mosquito [73, 79].

The variation in the amount of ingredient of curcuminoids and geranium oil affected the loading capacity and mean particle size of nanoformulation [71, 77]; they reported that 5.0% (w/w) stearic acid was found to be an optimum concentration for the formulation of solid lipid nanoparticles (SLNs). On the other hand, [78] showed that the oil-loading efficiency could reach 80% at the optimal ratio of garlic essential oil to 10% of polyethylene glycol (PEG) which was used as coated nanoparticles for the oil. The morphology and size of nanoparticles showed a round appearance and good dispersion, and its size was <240 nm in average diameter, likewise [16], determined the polydispersity index (PDI, which measures the size of distribution of nanoparticles) and loading efficiency for eight essential oil nanoparticles, and illustrated that the 10% ratio EO-polyethylene glycol showed the best relationship between a low polydispersion, narrow size distribution, and a high essential oil-loading efficiency; these nanoparticles had the biggest size in average diameter < 235 nm and a loading efficiency of >75%. In [71, 79], it is mentioned that the potential of the particles depending on the nature of the lipid matrix produced, which were used as coating nanoparticles, also influenced the encapsulation efficiency and accordingly loading capacity. In contrast, in [80], it is indicated that starch-coated encapsulation of neem oil nanoemulsion was found to be effective when compared to polyethylene glycol-coated encapsulation of neem oil nanoparticles.

The biological efficacy of geranium essential oil alone and in the form of nanoformulation was evaluated and compared against the potato tuber moth *Ph. operculella* first larval instar. This study showed that geranium oil-loaded solid lipid nanoparticles at different concentrations under laboratory conditions significantly affected the developmental process of immature stages and increased the percentage of mortality at all treatments and significantly reduced the adult's progeny and female fecundity and accordingly the percentage of hatchability. When this nanoformulated oil (geranium-loaded solid lipid nanoparticles) was applied under field conditions on potato crop, it exhibited longer residual efficacy than the free essential oil, suggesting that it may help to reduce insecticide application to control *Ph. operculella* [81].

It is known that nanoencapsulated oils have a much higher chemical activity than the bulk material, much more mobile, enabling penetration into insect tissues and enhancing insecticidal activity; this can be achieved by direct contact through the insects' cuticle or by ingestion and penetration through the digestive tract. They penetrate the cells of the pest especially epithelial and endothelial cells by transcytosis as confirmed by [76]. Nevertheless [79] concluded that the repellent effect of the obtained nanoemulsions composed of citronella oil, hairy basil, and vetiver oil could be attributed to the major difference in oil droplet size. The small nanoscale of droplet size of nanoemulsion prepared with high-pressure homogenization would play an important role on their efficacy besides physical stability. The prolonged mosquito protection time is probably due to the combination of these three essential oils. The lethal and sublethal activity of citrus peel essential oil as an emulsion and loading into polyethylene glycol nanoparticles was studied against the invasive tomato pest *Tuta absoluta* [82]. Their results showed that the essential oil nanoformulation tested had a significant insecticidal activity with higher mortality and significantly reduced the visible toxic effects on the plants suggesting that nanoformulated natural products could be successfully used in integrated pest management programs for *T. absoluta*. The insecticidal activity of *Rosmarinus officinalis* essential oil was enhanced in study by [83] for effective management of the red flour beetle, *Tribolium castaneum*, using nanoprecipitation method to prepare rosemary oil-loaded nanocapsules. In a research study done by [84], the contact toxicity of *Mentha longifolia* L. essential oil compared with its nanoemulsion on *Ephestia kuehniella* Zeller has been investigated; their results showed that the nanoemulsion formulation increased the effect of essential oil contact toxicity and its durability. The essential oil nanoformulations characterized by distinctive slow release property may represent a new category of biopesticide formulations that should be considered as a promising agent in the integrated pest management program.

5. Conclusion

The global population has been expanding rapidly for many years; in developing countries, it is expected that the food demand in 2050 will increase by 50–100% [85]; for this reason there is an urgent need to find safe alternative strategies that may contribute to the provision of food and at the same time protect the environment and human beings. To limit crop yield losses and increase the agricultural productivity, integrated pest management programs have implanted the application of effective, environmentally safe biopesticides.

Plant essential oils over the years were used as biopesticides to control insects; however, the difficulty in applying essential oils on large-scale and under severe environmental conditions required incorporation of these plant materials into new formulations through nanotechnology, such as nanoformulations that enhance the efficacy, increase stability, and prevent rapid evaporation of active compounds in plant oils. It is known that nanoencapsulated oils have a much higher chemical activity than the bulk material, are much more mobile, and are able to penetrate into insect tissues for efficient insecticidal activity.

The essential oil nanoformulations appear to be promising candidates to control the major pests of plants, due to their high volatility and stability. Before implementing the use of such oils, large-scale experiments are needed to evaluate their mammalian toxicity and to substantiate their efficacy under different conditions to validate their economic values as plant protectant; these nanoformulations require more in-depth studies to encourage the use of these natural substances in IPM programs and promote the development of sustainable crop-based systems that enhance crop yields, reduce ecological damage, and improve the quality of life for producers and consumers.

Conflict of interest

The author reports no conflicts of interest in this work.

Author details

Samar S. Ibrahim
Department of Pests and Plant Protection, National Research Centre, Cairo, Egypt

*Address all correspondence to: samarsayed66@yahoo.com

References

[1] Lindquist RK, Cloyd RA. Identification of Insects & Related Pests of Horticultural Plants: A Pictorial Guide. Columbus, Ohio: Ohio Florist Association; 2005

[2] Burfield T, Reekie SL. Mosquitoes, malaria and essential oils. International Journal of Aromatherapy. 2005;**15**:30-41

[3] El-Wakeil NE. Botanical pesticides and their mode of action (Review article). Gesunde Pflanzen. 2013;**65**: 125-149

[4] Ghormade V, Deshpande MV, Paknicar KM. Perspectives for nano biotechnology enabled protection and nutrition of plants. Biotechnology Advances. 2011;**29**(6):792-803

[5] Bertolini AC, Siani AC, Grosso CRF. Stability of monoterpenes encapsulated in gum Arabic by spray-drying. Journal of Agricultural and Food Chemistry. 2001;**49**:780-785

[6] Lai F, Sylvia AW, Rainer HM, Anna MF. Artemisia *arborescens L*. essential oil–loaded solid lipid nanoparticles for potential agricultural application: Preparation and characterization. AAPS PharmSciTech. 2006;7(1):E10-E18

[7] Rajendran S, Sriranjini V. Plant products as fumigants for stored-product insect control. Journal of Stored Products Research. 2008;**44**: 126-135

[8] Wink M. Production and application of phytochemicals from an agricultural perspective. In: van Beek TA, Breteler H, editors. Phytochemistry aned Agriculture. Vol. 34. Oxford, UK: Clarendon; 1993. pp. 171-213

[9] Bakkali F, Averbeck S, Averbeck D, Idaomar M. Biological effects of essential oils—A review. Food and Chemical Toxicology. 2008;**46**:446-475

[10] Chamorro ER, Zambón SN, Morales WG, Sequeira AF, Velasco GA. Study of the chemical composition of essential oils by gas chromatography. In: Salih B, editor. Gas Chromatography in Plant Science, Wine Technology, Toxicology and Some Specific Applications. Rijeka, Croatia: InTech; 2012. pp. 307-324. ISBN: 978-953- 51-0127-7

[11] Somesh M, Rupali S, Swati S, Jose M, Manish M. In-vitro comparative study on antimicrobial activity of five extract of few citrus fruit: Peel peel & pulp vs gentamicin. Australian Journal of Basic and Applied Sciences. 2015;**9**(1): 165-173

[12] Hesham HAR, Abdurahman HN, Rosli MY. Techniques for extraction of essential oils from plants: A Review. Australian Journal of Basic and Applied Sciences. 2016;**10**(16):117-127

[13] Bowers WS, Ohta T, Cleere JS, Marsella PA. Discovery of insect antijuvenile hormones in plants yield a potential fourth generation insecticides. Science. 1972;**193**:542-547

[14] Su HCF, Harvort R. Isolation identification and insecticidal properties of *Piper nigrumamides*. Journal of Agricultural and Food Chemistry. 1981; **29**:115-118

[15] Meisner J, Fleischerand A, Elzick C. Phagodeterrency induced by (−) Carvon in the larvae of *Spodoptera littoralis* (Boisd). Journal of Economic Entomology. 1982;**75**:462-466

[16] Werdin-Gonzalez JO, Gutiérrez MM, Ferrero AA, Band BF. Essential oils nanoformulations for stored-product pest control—Characterization and biological properties. Chemosphere. 2014;**100**:130-138

[17] Agostini-Costa TS, Vieira RF, Bizzo HR, Silveira D, Gimenes MA.

Chromatography and its applications. In: Dhanarasu S, editor. Plant Secondary Metabolites. Rijeka, Croatia: In Tech Publisher; 2012. pp. 131-164

[18] Bowsher C, Steer M, Tobin A. Plant Biochemistry. New York-Abigdon: Garland Science, Taylor and Francis Group. 2008. 446 pp. ISBN-0-8153-4121-0

[19] Lee BH, Annis PC, Tumaalii F, Choi W. Fumigant toxicity of essential oils from Myrtaceae family and 1, 8-cineol against 3 major stord—Grain insects. Journal of Stored Products Research. 2004;**40**:553-564

[20] Sharaby A. Effect of orange, *Citrus sinensis L.* peel oil on reproduction in *Phthorimaea operculella* (Zell.). Insect Science and Its Application. 1988;**9**(2): 201-203

[21] Al-Dhafer ZMA. Studies on the effect of natural products of some Saudi Arabian plants on the biology and behavior of the American bollworm *Helicovirpa armigera* (Hubn.) (Lepidoptera: Noctoidae) [thesis]. Fac. Sci; Girls College, Dammam, Saudi Arabia. 2001

[22] Al-Dosary MM. Sensory receptors and behaviour of the Red Palm Weevil *Rhynchophorus ferrugineus* (Oliv.) (Colioptera: Curculionidae) with reference to attractants, repellents and control [thesis]. Riyadh Girls College of Education Scientific Section, Department of Zoology. 2007

[23] Ozols G, Bicevskis M. Respects for the use of Ips tyrographus attractant. In: Shumakov EM, Chekmenev SY, Ivanova TV, editors. Biologia Aktualis Veshchestva Zashchiva Rastenij. Moscow: Izd. Kolos; 1979. pp. 49-51

[24] Mikhaiel AA. Potential of some volatile oils in protecting packages of irradiated wheat flour against *Ephestia kuheniella* and *Tribolium castaneum*. Journal of Stored Products Research. 2011;**47**(4):357-364. DOI: 10.1016/j.jspr.2011.06.002

[25] Mann RS, Kaufman PE. Natural product pesticides: Their development, delivery and use against insect vectors. Mini-Reviews in Organic Chemistry. 2012;**9**:185-202. DOI: 10.2174/157019312800604733

[26] Sharaby A, El-Nujiban A. Evaluation of some plant essential oils against the black cutworm *Agrotis ipsilon*. Global Journal of Advanced Research. 2015;**2**(4):701-711

[27] Huang Y, Ho SH, Lee HC, Yap YL. Insecticidal properties of eugenol, isoeugenol and methyleugenol and their effects on nutrition of *Sitophilus zeamais* Motsch. (Coleoptera: Curculionidae) and *Tribolium castaneum* (Herbst) (Coleoptera: Tenebrionidae). Journal of Stored Products Research. 2002;**38**: 403-412

[28] Schoonhoven AV. The use of vegetable oils to protect stored beans from bruchid attack. Journal of Economic Entomology. 1978;**71**(2): 254-256. DOI: 10.1093/jee/71.2.254

[29] Adedire CO, Obembe OM, Akinkurolere RO, Oduleye SO. Response of *Callosobruchus maculatus* Fabricius (Coleoptera: Chrysomelidae: Bruchinae) to extracts of cashew kernels. Journal of Plant Diseases Protection. 2011;**118**(2):75-79. DOI: 10.1007/BF03356385

[30] Ileke KD, Olotuah OF. Bioactivity of *Anacardium occidentals* and *Allium sativum* powders and oils extracts against cowpea bruchid, *Callosobruchus maculatus* (Fab) (Coleoptera: Bruchidae). International Journal of Biological Sciences. 2012;**4**(1):96-103

[31] Kaufmann C, Briegel H. Flight performance of the malaria vectors *Anopheles gambiae* and *Anopheles*

atroparous. Journal of Vector Ecology. 2004;**29**(1):140-153

[32] Hollingworth RM, Johnstone EM, Wright N. Pesticide synthesis through rational approaches. In: Magee PS, Kohn GK, Menn JJ, editors. ACS Symposium Series. Vol. 255. Washington, DC: American Chemical Society; 1984. pp. 103-125

[33] Enan EE. Molecular and pharmacological analysis of an octopamine receptor from american cockroach and fruit fly in response to plant essential oils. Archives of Insect Biochemistry and Physiology. 2005;**59**: 161-171. DOI: 10.1002/(ISSN)1520-6327

[34] Souguir S, Chaieb I, Ben Cheikh Z, Laarif A. Insecticidal activities of essential oils from some cultivated aromatic plants against *Spodoptera littoralis* (Boisd.). Journal of Plant Protection Research. 2013;**53**(4):388-391

[35] Schoonhoven L. Biological aspects of antifeedant. Entomologia Experimentalis et Applicata. 1982;**31**: 57-69

[36] Scriber JM, Slansky F. The nutritional ecology of immature insects. Annual Review of Entomology. 1981;**26**: 183-211

[37] Reese JC, Chan BG, Waiss AC. Effects of cotton condensed tannin, maysin (Corn) and pinitol (soybeans) on *Heliothis zea* growth and development. Journal of Chemical Ecology. 1982;**8**(12):1429-1436

[38] Sharaby A, Abdel Rahman H, Abdel-Aziz SS, Moawad SS. Natural plant oils and terpenes as protector for the potato tubers against *Phthorimaea operculella* infestation by different application methods. Ecologia Balkanica. 2014;**6**(1):45-59

[39] Zapata N, Budia F, Viñuela E, Medina P. Antifeedant and growth inhibitory effects of extracts and drimanes of Drimys winteri stem bark against *Spodoptera littoralis* (Lep., Noctuidae). Industrial Crops and Products. 2009;**30**:119-125

[40] Stroh J, Wan MT, Isman MB, Moul DJ. Evaluation of the acute toxicity to juvenile Pacific coho salmon and rainbow trout of some plant essential oils, a formulated product, and the carrier. Bulletin of Environmental Contamination and Toxicology. 1998; **60**:923-930

[41] Koul O, Singh R, Kaur B, Kanda D. Comparative study on the behavioral response and acute toxicity of some essential oil compounds and their binary mixtures to larvae of *Helicoverpa armigera*, *Spodoptera litura* and *Chilo partellus*. Industrial Crops and Products. 2013;**49**:428-436

[42] Gutiérrez L, Escudero A, Batlle R, Nerín C. The effect of mixed antimicrobial agents and flavours in active packaging films. Journal of Agricultural and Food Chemistry. 2009; **57**:8564-8571

[43] Calvache H. Efecto de barreras vegetales y quı'micas en el control del gusano blanco de la papa (*Premnotrypes vorax*). Revista Latinoamericana de la Papa. 1991;**4**:22-35

[44] Kroschel J, Koch W. Studies on the use of chemicals, botanicals and *Bacillus thuringiensis* in the management of the potato tuber moth in potato stores. Crop Protection. 1996;**15**(2):197-203

[45] Go'mez Jime'neza MI, Poveda K. Synergistic effects of repellents and attractants in potato tuber moth control. Basic and Applied Ecology. 2009;**10**: 763-769

[46] Labinas AM, Crocomo WB. Effect of Java grass (*Cymbopogon winterianus* Jowitt) essential oil on fall armyworm *Spodoptera frugiperda* (J. E. Smith, 1797)

(Lepidoptera, Noctuidae). Maringá. 2002;**24**(5):1401-1405

[47] Jeyasankar A. Antifeedant, insecticidal and growth inhibitory activities of selected plant oils on black cutworm, *Agrotis ipsilon* (Hufnagel) (Lepidoptera: Noctuidae). Asian Pacific Journal of Tropical Disease. 2012;**2**(1): S347-S351

[48] Ho SH, Koh L, Ma Y, Huang Y, Sim KY. The oil of garlic, *Allium sativum* L. (Amaryllidaceae), as a potential grain protectant against *Tribolium castaneum* (Herbst) and *Sitophilus zeamais* Motsch. Postharvest Biology and Technology. 1996;**9**:41-48

[49] Kosti'ca I, Petrovi'c O, Milanovi'c S, Popovi'c Z, Stankovi'c S, Todorovi'c G, et al. Biological activity of essential oils of *Athamanta haynaldii* and *Myristica fragrans* to gypsy moth larvae. Industrial Crops and Products. 2013;**41**:17-20

[50] Pinheiro PF, de Queiroz VT, Rondelli VM, Costa AV, Marcelino TP, Pratissoli D. Insecticidal activity of citronella grass essential oil on Frankliniella schultzei and Myzus persicae. Ciência e Agrotecnologia, Lavras. 2013;**37**(2):138-144

[51] Papachristos DP, Kimbaris AC, Papadopoulos NT, Polissiou MG. Toxicity of citrus essential oils against *Ceratitis capitata* (Diptera: Tephritidae) larvae. The Annals of Applied Biology. 2009;**155**:381-389

[52] Hussein HS, Salem ZMM, Soliman MA. Repellent, attractive, and insecticidal effects of essential oils from *Schinus terebinthifolius* fruits and *Corymbia citriodora* leaves on two whitefly species, *Bemisia tabaci*, and *Trialeurodes ricini*. Scientia Horticulturae. 2017;**216**:111-119

[53] Stephenson GR. Pesticide use and world food production: Risks and benefits. Environmental Fate and Effects of Pesticides. 2003:261-270. DOI: 10.1021/bk-2003-0853.ch015

[54] Tripathi AK, Upadhyay S, Bhuiyan M, Bhattacharya PR. A review on prospects of essential oils as biopesticide in insect-pest management. Journal of Pharmacognosy and Phytotherapy. 2009;**15**:52-63

[55] Isman MB. Plant essential oils for pest and disease management. Crop Protection. 2000;**19**:603-608

[56] Moretti MDL, Sanna-Passino G, Demontis S, Bazzoni E. Essential oil formulations useful as a new tool for insect pest control. AAPS PharmSciTech. 2002;**3**(2):1-11

[57] Frederiksen HK, Kristensen HG, Pedersen M. Solid-lipid nanoparticle formulations (SLN) of the pyrethroid gamma-cylalothrin (GCH): incompatibility of the lipid and the pyrethroid. Journal of Controlled Release. 2003;**86**:243-253

[58] Liu WT. Nanoparticles and their biological and environmental applications. Journal of Bioscience and Bioengineering. 2006;**102**:1-7

[59] McClements DJ. Nanoemulsions versus microemulsions: Terminology, differences, and similarities. Soft Matter. 2012;**8**:1719-1729

[60] McClements DJ. Food Emulsions: Principles, Practices, and Techniques. Boca Raton, London: CRC Press; 2004. ISBN: 0849320232

[61] Anton N, Benoit JP, Saulnier P. Design and production of nanoparticles formulated from nano-emulsion templates—A review. Journal of Controlled Release. 2008;**128**:185-199

[62] Vilasau J, Solans C, Gómez M, Dabrio J, Mújika-Garai R, Esquena J. Phase behaviour of a mixed ionic/nonionic surfactant system used to

prepare stable oil-in-water paraffin emulsions. Colloids and Surfaces A: Physicochemical and Engineering Aspects. 2011;**384**:473-481

[63] Nagpal BN, Srivastava A, Valecha NA, Sharma VP. Repellent action of neem cream against *An. culicifacies* and *Cx. Quinquefasciatus*. Current Science. 2001;**80**:1270-1271

[64] Kumari A, Yadav SK, Yadav SC. Biodegradable polymeric nanoparticles based drug delivery systems. Colloids and Surfaces. 2010;**75**(1):1-18

[65] Reis CP, Neufeld RJ, Ribeiro AJ, Veiga F. Nanoencapsulation I. Methods for preparation of drug-loaded polymeric nanoparticles. Nanomedicine: Nanotechnology, Biology and Medicine. 2006;**2**:8-21

[66] Bangham AD. Physical structure and behavior of lipids and lipid enzymes. Advances in Lipid Research. 1963;**1**:65-104

[67] Batzri S, Korn ED. Single bilayer liposomes prepared without sonication. Biochimica et Biophysica Acta. 1973; **298**:1015-1019

[68] Pardeike J, Hommoss A, Müller RH. Lipid nanoparticles (SLN, NLC) in cosmetic and pharmaceutical dermal products. International Journal of Pharmaceutics. 2009;**366**:170-184

[69] São Pedro A, Santo IE, Silva C, Detoni C, Albuquerque E. The use of nanotechnology as an approach for essential oil-based formulations with antimicrobial activity. In: Méndez-Vilas A, editor. Microbial Pathogens and Strategies for Combating them: Science, Technology and Education. Zurbaran, Badajoz, Spain: Formatex Research Center; 2013. pp. 1364-1374

[70] Nel A, Xia T, Madler L, Li N. Toxic potential of materials at the nanolevel. Science. 2006;**311**(5761):622-627

[71] Nayak AP, Tiyaboonchai W, Patankar S, Madhusudhana B, Souto EB. Curcuminoids-loaded lipid nanoparticles: Novel approach towards malaria treatment. Colloids and Surfaces B: Biointerfaces. 2010;**81**: 263-273

[72] Kaur IP, Bhandari R, Bhandari S, Kakkar V. Potential of solid lipid nanoparticles in brain targeting. Journal of Controlled Release. 2008; **127**:97-109

[73] Solomon B, Sahle FF, Gebre-Mariam T, Asres K, Neubert RHH. Microencapsulation of citronella oil for mosquito-repellent application: Formulation and in vitro permeation studies. European Journal of Pharmaceutics and Biopharmaceutics. 2012;**80**:61-66

[74] Anjali CH, Sudheer Khan S, Margulis-Goshen K, Magdassi S, Mukherjee A, Chandrasekaran N. Formulation of water-dispersible nanopermethrin for larvicidal applications. Ecotoxicology and Environmental Safety. 2010;**73**: 1932-1936

[75] Khot L, Sankaran S, Maja J, Ehsani R, Schuster E. Applications of nanomaterials in agricultural production and crop protection: A review. Crop Protection. 2012;**35**:64-70

[76] Devi N, Maji TK. Study of complex coacervation of gelatin A with sodium carboxymethyl cellulose: Microencapsulation of neem (*Azadirachta indica* A. Juss.) seed oil (NSO). International Journal of Polymeric Materials and Polymeric Biomaterials. 2011;**60**(13):1091-1105

[77] Asnawi S, Abd Aziz A, Abd Aziz R, Khamis AK. Formulation of geranium oil loaded solid lipid nanoparticles for mosquito repellent application. Journal of Chemical and Natural Resources Engineering. 2008;**2**:90-99

[78] Yang FL, Li SG, Zhu F, Lei CL. Structural characterization of nanoparticles loaded with garlic essential oil and their insecticidal activity against *Tribolium castaneum* (Herbst) (Coleoptera: Tenebrionidae). Journal of Agricultural and Food Chemistry. 2009;**57**:10156-10162

[79] Nuchuchua O, Sakulku U, Uawongyart N, Puttipipatkhachorn S, Soottitantawat A, Ruktanonchai1 U. In vitro characterization and mosquito (*Aedes aegypti*) repellent activity of essential-oils-loaded nanoemulsions. AAPS PharmSciTech. 2009;**10**(4): 1234-1242

[80] Jerobin J, Sureshkumar RS, Anjali CH, Mukherjee A, Chandrasekaran N. Biodegradable polymer based encapsulation of neem oil nanoemulsion for controlled release of Aza-A. Carbohydrate Polymers. 2012;**90**(4): 1750-1756

[81] Ibrahim SS. Applications of some biotechnological methods for controlling the most important pests of potatoes [thesis]. Fac. Sci, Al-Azhar University, Egypt. 2017

[82] Campolo O, Cherif A, Ricupero M, Siscaro G, Grissa-Lebdi K, Russo A, et al. Citrus peel essential oil nanoformulations to control the tomato borer, *Tuta absoluta*: Chemical properties and biological activity. Scientific Reports. 2017;7:13036. DOI: 10.1038/s41598-017-13413-0

[83] Khoobdel M, ahsaei SA, farzaneh M. Insecticidal activity of polycaprolactone nanocapsules loaded with *Rosmarinus officinalis* essential oil in *Tribolium castaneum* (Herbst). Entomological Research. 2017;**47**:175-184

[84] Louni M, Shakarami J, Negahban M. Insecticidal efficacy of nanoemulsion containing *Mentha longifolia* essential oil against *Ephestia kuehniella* (Lepidoptera: Pyralidae). Journal of Crop Protection. 2018;7(2):171-182

[85] Alexandratos N, Bruinsma J. World Agriculture Towards 2030/2050: The 2012 Revision. ESA Working Paper; Rome: FAO; 2012